FOUNDATIONS OF COLLEGE CHEMISTRY IN THE LABORATORY

Thirteenth Edition

Selected Experiments

Morris Hein
Judith N. Peisen
Robert L. Miner

Moraine Valley Community College

The author of this manual has outlined extensive safety precautions in each experiment. Ultimately, it is your responsibility to follow safe laboratory guidelines and procedures. The author and publisher disclaim any liability for any loss or damage claimed to have resulted from or been related to the experiments.

To order books or for customer service, please call 1(800)-CALL-WILEY (225-5945).

ISBN 978-1-118-33089-0

10 9 8 7 6 5 4 3

FOUNDATIONS OF CHEMISTRY IN THE LABORATORY

THIRTEENTH EDITION

Morris Hein
Mount San Antonio College

Judith N. Peisen
Hagerstown Community College

Robert L. Miner
Mount San Antonio College

John Wiley & Sons, Inc.

To order books or for customer service, please call 1-800-CALL-WILEY (225-5945).

ISBN-13   978-0-470-55490-6

Printed in the United States of America

10 9 8 7 6 5 4 3 2 1

Printed and bound by Bind-Rite, Inc.

CONTENTS

Preface

This manual is intended for the student who has not had a course in chemistry. The experiments are designed to be challenging but understandable to the student. Experimentation begins with simple laboratory techniques and measurements and progresses to relatively complex procedures. The 28 experiments are graded in difficulty to keep pace with the expanding capability of the student. The number and variety of experiments allow the instructor reasonable flexibility in preparing a laboratory schedule taht expands and supports many lecture topics in a one- or two-semester preparatory college chemistry course.

Our major objectives of this flexible laboratory program are to provide experience with (1) hands-on laboratory experimentation, (2) the capabilities and limitations of measurements, (3) a variety of chemical reactions and the equations used to describe them, (4) the collection, analysis, and graphing of data, (5) responsible disposal of chemicals for personal and environmental health, (6) Using a computer for graphing of data, (7) drawing valid conclusions form experimental evidence, and (8) support and reinforcement of concepts introduced in the lecture component of the course.

We have tried to maintain a balance between descriptive and quantitative experiments. All the experiments have been reviewed and modified to eliminate the use of heavy metals (where possible) and minimize exposure to hazardous material. Five experiments include unknowns for student analysis, and seven provide opportunities for graphing data. The Instructor's Manual provides sample student data, including graphs, for every experiment. The Instructor's Manual is available on the textbook website.

The format is designed to be helpful and convenient for both student and instructor and includes the following features:

1. A concise discussion of the basic underlying principles for each experiment provides pertinent background material to supplement, not replace, the textbook.

2. Six Study Aids provide supplementary material common to several experiments on the important topics of (a) significant figures, (b) formulas and chemical equations, (c) reading and preparing graphs by hand and by computer, (d) using a scientific calculator, (e) dimensional analysis and stoichiometry, and (f) introduction to organic chemistry.

3. Experimental procedures have been extensively tested by many students and provide enough detail for students to work with only general supervision.

4. Report forms for each experiment are cross-referenced to letters and subtitles in the procedure, designed to be completed before leaving the lab session, and relatively easy to grade.

5. The names and formulas of reagents used are listed at the beginning of each experiment.

6. Special safety precautions and waste disposal instructions are indicated when necessary at the point where they are required within the procedure.

7. For the convenience of the instructor and stockroom personnel, the appendices provide (a) an experiment-by-experiment list of special equipment and preparations needed, (b) a list of suggested equipment for student lockers, (c) an experiment-by-experiment list of waste disposal instructions, (d) a list of suggested auxiliary equipment, and (e) a complete list of reagents and details for the preparation of solutions.

The experiment which is new to the Thirteenth Edition is Experiment 8, Water, Solutions, and pH. This experiment provides an introduction to the properties of water especially important in the life sciences, introduces some skills used in biotechnology and the concept of molarity. This new experiment is an excellent foundation for Experiment 9, Properties of Solutions which is found in previous editions. Properties of Lead (II), Silver, and Mercury (I) Ions (Experiment 14 in the twelfth edition) which involved the use of heavy metal cations, has been eliminated.

Meticulous instructions for waste disposal have been continued and updated for students within each procedure and in the Instructor's Manual. The instructions for Preparing a Graph (Study Aid 3) have been updated to the most recent version of Excel (2007).

We are especially indebted to students in the chemistry departments of Mount San Antonio College and Hagerstown Community College for their patience and helpful suggestions during the development and testing of this laboratory program. We appreciate the feedback from instructors and students at the many schools over the years that have used this lab manual in their introductory chemistry course. A special thanks to Dr. Richard Montgomery, Dr. William Elliott, and Dr. Melanie Ulrich for their contribution to the Water, Solutions, and pH experiment that is new to this edition. Further suggestions for improvements of material in this laboratory manual are always welcome.

M. Hein
J. N. Peisen
R. L. Miner

To the Student

Since your laboratory time is limited, it is important to come to each session prepared by at least one hour of detailed study of the scheduled experiment. This should be considered a standing homework assignment.

Each of the experiments in this manual is composed of four parts:

1. **Materials and Equipment**—a list that includes the formulas of all substances used.

2. **Discussion**—a brief discussion of the principles underlying the experiment.

3. **Procedure**—detailed directions for performing the experiment with safety precautions clearly noted and disposal procedures for chemical waste provided throughout and identified by a waste icon.

4. **Report for Experiment**—a form for recording data and observations, performing calculations, and answering questions.

Follow the directions in the procedure carefully, and consult your instructor if you have any questions. For convenience, the letters and subtitles in the report form have been set up to correspond with those in the procedure section of each experiment.

As you make your observations and obtain your data, record them on the report form. Try to use your time efficiently; when a reaction or process is occurring that takes considerable time and requires little watching, start working on other parts of the experiment, perform calculations, answer questions on the report form, or clean up your equipment.

Except when your instructor directs otherwise, you should do all the work individually. You may profit by discussing experimental results with your classmates, but in the final analysis you must rely on your own judgment in completing the report form.

⚠ Safety Guidelines

While in the chemistry laboratory, you are responsible not only for your own safety but for the safety of everyone else. *We have included safety precautions in every experiment where needed, and they are highlighted with the icon shown in the title of this section.* Your instructor may modify these instructions and give you more specific directions on safety in your laboratory. If the proper precautions and techniques are used, none of the experiments in this laboratory program are hazardous. But without your reading and following the instructions, without knowledge about handling and disposal of chemicals, and without the use of common sense at all times, accidents can happen. Even when everyone is doing his or her best to comply with the safety guidelines in each experiment, accidents can happen. It is your responsibility to minimize these accidents and know what to do if they happen.

Laboratory Rules and Safety Procedures

1. **Wear protective goggles or glasses** at all times in the laboratory work area. These glasses should wrap around the face so liquids cannot splash into the eye from the side. These goggles are mandated by eye-protection laws and are not optional, even though they maybe uncomfortable. Contact lenses increase the risk of problems with eye safety, even when protective goggles are worn. If you wear contact lenses, inform the instructor.

2. **Dress appropriately** for the laboratory. Shoes that do not completely cover the foot are not allowed (*no sandals*). Long hair should be tied back. Wear a laboratory coat or apron, if available, to protect your clothing.

3. **Keep your benchtop organized as you work.** Put jackets, book bags, and personal belongings away from the work areas. Before you leave, clean your work area and make sure the gas and water are turned off. Clean and return all glassware and equipment to your drawer or the lab bench where you borrowed it.

4. **Keep all stock bottles of solid and liquid reagents in the dispensing area.** Do not bring reagent bottles to your laboratory work area. Use test tubes, beakers, or weigh boats to obtain chemicals from the dispensing areas: (1) the reagent shelf, (2) the balance tables, (3) under the fume hood, and (4) as instructed.

5. **Keep the balance and the area around it clean.** Do not place chemicals directly on the balance pans; place a piece of weighing paper, a weigh boat, or another small container on the pan first, and then weigh your material. **Never weigh an object while it is hot.**

6. **Carefully check the name on the reagent bottles before you use them.** Many names and formulas appear similar at first glance. Label every beaker, test tube, etc., into which you transfer chemicals. Many labels will contain the National Fire Protection Association (NFPA) diamond label, which provides information about the flammability, reactivity, health effects, and miscellaneous effects for the substance. Each hazard is rated 0 (least hazardous) to 4 (most hazardous). For example, the NFPA label for potassium chromate is shown below.

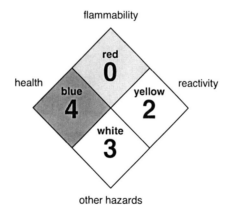

More specific information (the reason for potassium chromate being rated an extreme health hazard, for example) about all known substances is available in the form of Material Safety Data Sheets (MSDS), which many institutions keep on file for chemicals stored and used in their laboratories. MSDS are usually provided with chemicals by the supplier when they are purchased and are easily obtained from many website sources. Because of its hazardous nature, chromates have been removed from this lab manual.

7. **Never return unused chemicals to the reagent bottles.** This is a source of possible contamination of the entire stock bottle. Dispose of unused chemicals exactly as instructed in the waste disposal instructions for that substance, identified by throughout each experiment.

8. **Disposal of wastes must follow state and federal guidelines.** Do not put anything into the trash or sink without thinking first. We have tried to anticipate every disposal decision in the procedure and marked the procedure with the waste icon. The following guidelines are the foundation of waste disposal decisions:

    a. Broken glass is put into a clearly marked special container.

    b. Organic solvents are never poured into the sink. They are usually flammable and often immiscible with water. Instead, they are poured into a specially marked container ("waste organic solvents") provided when needed.

    c. Solutions containing cations and anions considered toxic by the EPA are never poured into the sink. They are poured into specially marked containers ("waste heavy metal," etc.) provided when needed. The name of all ions disposed of into a specific bottle must be listed on the label.

    d. Solutions poured in the sink should be washed down with plenty of water.

    e. Some solid chemicals must also be disposed of in specially labeled containers. If you are not sure what to do, ask the instructor.

    f. Each school may have its own policy for waste disposal which supercedes the instructions in this manual.

9. **Avoid contaminating stock solutions.** Do not insert medicine droppers or pipets into reagent bottles containing liquids. Instead, pour a little of the liquid into a small beaker or test tube. If the bottle is fitted with a special pipet that is stored with the bottle, this may not be necessary.

10. **Avoid all direct contact with chemicals.**

    a. Wash your hands anytime you get chemicals on them and at the end of the laboratory session.

    b. If you spill something, clean it up immediately before it dries or gets on your papers or skin.

    c. Never pipet by mouth.

    d. Never eat, drink, or smoke in the laboratory.

    e. Do not look down into the open end of a test tube in which a reaction is being conducted, and do not point the open end of a test tube at someone else.

    f. Inhale odors and chemicals with great caution. Waft vapors toward your nose. The fume hood will be used for all irritating and toxic vapors.

11. **Working with glass requires special precautions:**

    a. Do not heat graduated cylinders, burets, pipets, or bottles with a burner flame.

    b. Do not hold a test tube or beaker in your hand during a chemical reaction.

c. Do not touch glass that has been near a flame or hot plate. Hot glass looks the same as cool glass and may cause serious burns.

d. Learn and practice proper procedures when inserting glass tubing into rubber stoppers.

12. **Learn the location and proper use of safety equipment:** fire extinguisher, eye wash, first aid kit, fire blanket, safety shower, spill kits, and other equipment available.

13. **Never work alone** in the laboratory area.

14. **Report all accidents** to the instructor, no matter how minor.

15. **Do not perform unauthorized experiments.**

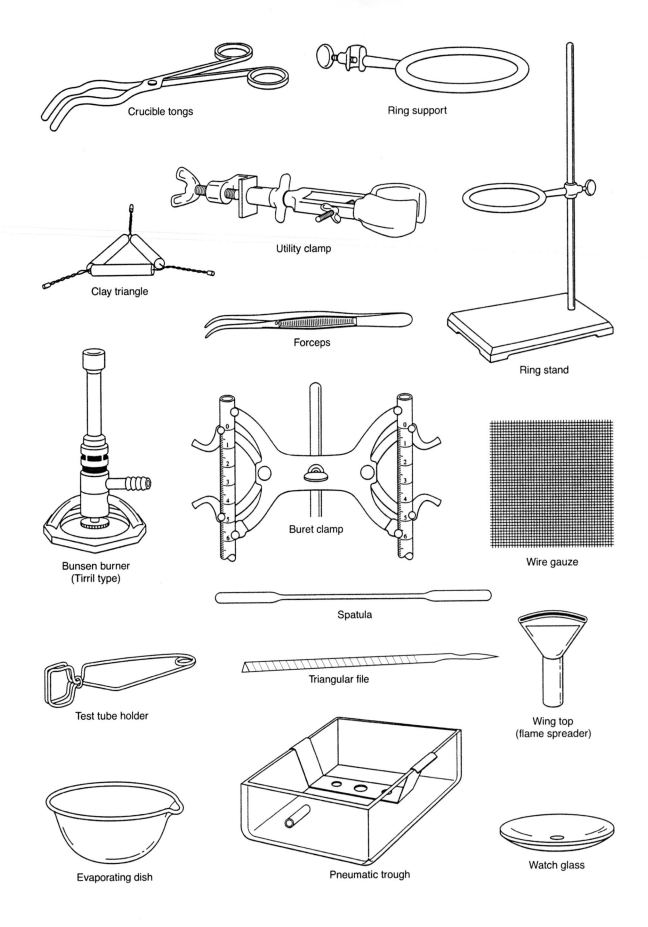

Crucible tongs

Ring support

Utility clamp

Clay triangle

Forceps

Ring stand

Bunsen burner
(Tirril type)

Buret clamp

Wire gauze

Spatula

Test tube holder

Triangular file

Wing top
(flame spreader)

Evaporating dish

Pneumatic trough

Watch glass

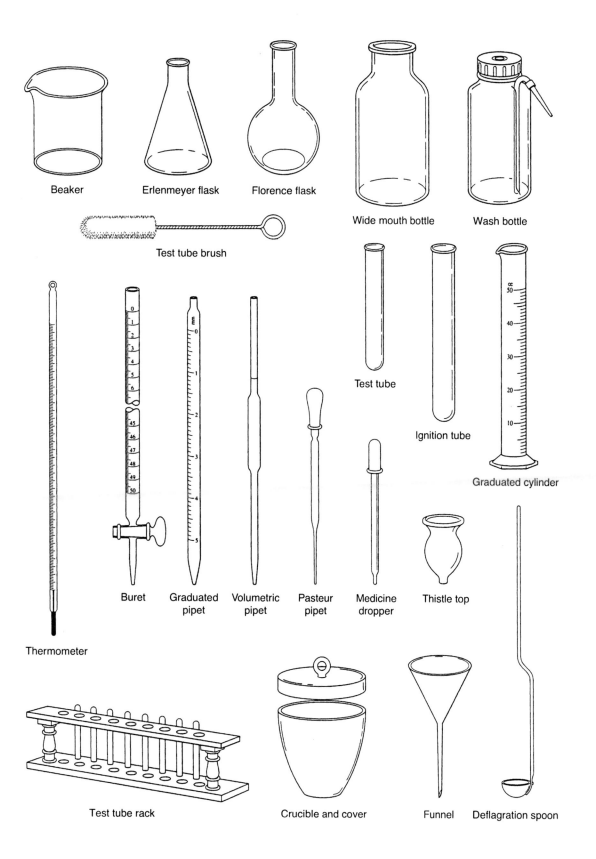

Beaker

Erlenmeyer flask

Florence flask

Wide mouth bottle

Wash bottle

Test tube brush

Test tube

Ignition tube

Graduated cylinder

Thermometer

Buret

Graduated pipet

Volumetric pipet

Pasteur pipet

Medicine dropper

Thistle top

Test tube rack

Crucible and cover

Funnel

Deflagration spoon

EXPERIMENT 2

Measurements

MATERIALS AND EQUIPMENT

Solids: sodium chloride (NaCl) and ice. Balance, ruler, thermometer, solid object for density determination, No. 1 or 2 solid rubber stopper.

DISCUSSION

Chemistry is an experimental science, and measurements are fundamental to most of the experiments. It is important to learn how to make and use these measurements properly.

The SI System of Units

The International System of Units (*Systeme Internationale, SI*) or metric system is a decimal system of units for measurements used almost exclusively in science. It is built around a set of units including the meter, the gram, and the liter and uses factors of 10 to express larger or smaller multiples of these units. To express larger or smaller units, prefixes are added to the names of the units. Deci, centi, and milli are units that are 1/10, 1/100, and 1/1000, respectively, of these units. The most common of these prefixes with their corresponding values expressed as decimals and powers of 10 are shown in the table below.

Prefix	Decimal Equivalent	Power of 10	Examples
deci (d)	0.1	10^{-1}	1 dg = 0.1 g = 10^{-1} g
centi (c)	0.01	10^{-2}	1 cm = 0.01 m = 10^{-2} m
milli (m)	0.001	10^{-3}	1 mg = 0.001 g = 10^{-3} g
kilo (k)	1000	10^{3}	1 km = 1000 m = 10^{3} m

Dimensional Analysis

It will often be necessary to convert from the American System of units to the SI system or to convert units within the SI system. Conversion factors are available from tables (see Appendix 4) or can be developed from the metric prefixes and their corresponding values as shown in the table above. Dimensional analysis, a problem-solving method with many applications in chemistry, is very valuable for converting one unit to another by the use of conversion factors. A review of using dimensional analysis for converting units is provided here. Study Aid 5 provides more help with this problem-solving tool.

Conversion Factors come from equivalent relationships, usually stated as equations. From each equivalence statement two conversion factors can be written in fractional form with a value of 1. For example:

Equivalence Equations	Conversion Factor #1	Conversion Factor #2
1 dollar = 4 quarters	$\dfrac{1 \text{ dollar}}{4 \text{ quarters}}$	$\dfrac{4 \text{ quarters}}{1 \text{ dollar}}$
1 lb = 453.6 g	$\dfrac{1 \text{ lb}}{453.6 \text{ g}}$	$\dfrac{453.6 \text{ g}}{1 \text{ lb}}$
1 mm = 10^{-3} m	$\dfrac{1 \text{ mm}}{10^{-3}\text{m}}$	$\dfrac{10^{-3}\text{m}}{1 \text{ mm}}$
1 ns = 10^{-9} s	$\dfrac{1 \text{ ns}}{10^{-9}\,\text{s}}$	$\dfrac{10^{-9}\,\text{s}}{1 \text{ ns}}$

The dimensional analysis method of converting units involves organizing one or more conversion factors into a logical series which cancels or eliminates all units except the unit(s) wanted in the answer.

For example: To convert 2.53 lb into milligrams (mg), the setup is:

$$(2.53 \text{ lb})\left(\frac{453.6 \text{ g}}{1 \text{ lb}}\right)\left(\frac{1 \text{ mg}}{10^{-3}\text{g}}\right) = 1.15 \times 10^{6} \text{ mg}$$

Note, that in completing this calculation, units are treated as numbers, **lb** in the denominator is canceled into **lb** in the numerator and **g** in the denominator is cancelled into **g** in the numerator. More examples of unit conversions can be found in Study Aid 5.

Although the SI unit of temperature is the Kelvin (K), the Celsius (or centigrade) temperature scale is commonly used in scientific work and the Fahrenheit scale is commonly used in this country. On the Celsius scale the freezing point of water is designated 0°C, the boiling point 100°C.

Precision and Accuracy of Measurements

Scientific measurements must be as **precise** as possible. This means that every measurement will include one uncertain or estimated digit. When making measurements we normally estimate between the smallest scale divisions on the instrument being used. Then, only the uncertain digit should vary if the measurement is repeated using the same instrument, even if it is repeated by someone else. The **accuracy** of a measurement or calculated quantity refers to its agreement with some known value. For example, we need to make two measurements, volume and mass, to determine the density of a metal. This experimental density can then be compared with the density of the metal listed in a reference such as the *Handbook of Chemistry and Physics*. High accuracy means there is good agreement between the experimental value and the known value listed in the reference. Not all measurements can be compared with a known value.

Random and Systematic Errors

The difference between the experimentally measured value of something and the accepted value of something is known as **the error.** For many of the experiments in this course, after you determine the error in your result, you may be required to find the percent error:

$$\text{Percent error} = \frac{\text{theoretical accepted value} - \text{experimentally determined value}}{\text{theoretical accepted value}} \times 100$$

There are two different types of error. **A random error** means that the error has an equal probablilty of being higher or lower than the accepted value. For example, a student measures the density of a quartz sample four times: (Accepted density value for quartz is 2.65 g/mL)

2.72 g/mL
2.55 g/mL Since two of the measured density values are below the mean
2.68 g/mL and two are above the mean, there is an **equal probability** of the
2.60 g/mL measurements being above or below the mean. This is a **random** error.
Since the mean density value is very close to the accepted value, the
Mean = 2.64 g/mL accuracy of the mean measurement is good. (the percent error is 0.38%)

The other type of error is a **systematic error.** This type of error occurs in the same direction each time (either always higher or always lower than the accepted value). For example, a student measures the boiling point of water four times (accepted temperature for the boiling point of water is 100.0° C.)

101.2° C
100.9° C Since all four of the measured temperature values are above the accepted
102.0° C value, the **error** is systematic. The mean value is 1.3% higher than the
101.0° C accepted value so the accuracy of these measurements is not as good as the
Mean = 101.3° C accuracy of the density of the measurements in the first example.

Precision and Significant Figures

When a measured value is determined to the highest precision of the measuring instrument, the digits in the measurement are called **significant digits** or **significant figures.**

Suppose we are measuring two pieces of wire, using the metric scale on a ruler that is calibrated in tenths of centimeters as shown in Figures 2.1a and b. One end of the first wire is placed at exactly 0.0 cm and the other end falls somewhere between 6.3 cm and 6.4 cm. Since the distance between 6.3 and 6.4 is very small, it is difficult to determine the next digit exactly. One person might estimate the length of the wire as 6.34 cm and another as 6.33 cm. The estimated digit is never ignored because it tells us that the ruler can be read to the 0.01 place. This measurement therefore has three significant figures (two certain and one uncertain figure).

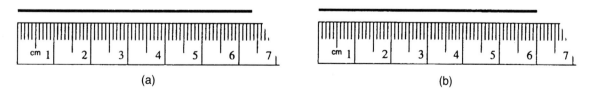

(a) (b)

Figure 2.1

The second wire has a length which measures exactly 6 cm on the ruler as shown in Figure 2.1b. Reporting this length as 6 cm would be a mistake for it would imply that the 6 is an uncertain digit and others might record 5 or 7 as the measurement. Recording the measurement as 6.0 would also be incorrect because it implies that the 0 is uncertain and that someone else might estimate the length as 6.1 or 5.9. What we really mean is that, as closely as we can read it, the length is exactly 6 cm. So, we must write the number in such a way that it tells how precisely we can read it. In this example we can estimate to 0.01 cm so the length should be reported as 6.00 cm.

Significant Figures in Calculations

The result of multiplication, division, or other mathematical manipulation cannot be more precise than the least precise measurement used in the calculation. For instance, suppose we have an object that weighs 3.62 lb and we want to calculate the mass in grams. $(3.62 \text{ lb})\left(\dfrac{453.6 \text{ g}}{1 \text{ lb}}\right) = 1{,}642.032$ when done by a calculator. To report 1,642.032 g as the mass is absurd, for it implies a precision far beyond that of the original measurement. Although the conversion factor has four significant figures, the mass in pounds has only three significant figures. Therefore the answer should have only three significant figures; that is, 1,640 g. In this case the zero cannot be considered significant. This value can be more properly expressed as 1.64×10^3 g. For a more comprehensive discussion of significant figures see Study Aid 1.

Precise Quantities versus Approximate Quantities

In conducting an experiment it is often unnecessary to measure an exact quantity of material. For instance, the directions might state, "Weigh about 2 g of sodium sulfite." This instruction indicates that the measured quantity of salt should be 2 g plus or minus a small quantity. In this example 1.8 to 2.2 g will satisfy these requirements. To weigh exactly 2.00 g or 2.000 g wastes time since the directions call for approximately 2 g.

Sometimes it is necessary to measure an amount of material precisely within a stated quantity range. Suppose the directions read, "weigh about 2 g of sodium sulfite to the nearest 0.001 g." This instruction does not imply that the amount is 2.000 g but that it should be between 1.8 and 2.2 g and measured and recorded to three decimal places. Therefore, four different students might weigh their samples and obtain 2.141 g, 2.034 g, 1.812 g, and 1.937 g, respectively, and each would have satisfactorily followed the directions.

Temperature

The simple act of measuring a temperature with a thermometer can easily involve errors. Not only does the calibration of the scale on the thermometer limit the precision of the measurement, but the improper placement of the thermometer bulb in the material being measured introduces a common source of human error. When measuring the temperature of a liquid, one can minimize this type of error by observing the following procedures:

1. Hold the thermometer away from the walls of the container.

2. Allow sufficient time for the thermometer to reach equilibrium with the liquid.

3. Be sure the liquid is adequately mixed.

When converting from degrees Celsius to Fahrenheit or vice versa, we make use of the following formulas:

$$°C = \frac{(°F - 32)}{1.8} \text{ or } °F = (1.8 \times °C) + 32$$

Example Problem: Convert 70.0°F to degrees Celsius:

$$°C = \left(\frac{70.0°F - 32}{1.8}\right) = \frac{38.0}{1.8} = 21.11°C \text{ rounded to } 21.1°C$$

This example shows not only how the formula is used but also a typical setup of the way chemistry problems should be written. It shows how the numbers are used, but does not show the multiplication and division, which should be worked out by calculator. The answer was changed from 21.11°C to 21.1°C because the initial temperature, 70.0°F, has only three significant figures. The 1.8 and 32 in the formulas are exact numbers and have no effect on the number of significant figures.

Mass (Weight)

The directions in this manual are written for a 0.001 gram precision balance, but all the experiments can be performed satisfactorily using a 0.01 gram or 0.0001 gram precision balance. Your instructor will give specific directions on how to use the balance, but the following precautions should be observed:

1. The balance should always be "zeroed" before anything is placed on the balance pan. On an electronic digital balance, this is done with the "tare" or "T" button. Balances without this feature should be adjusted by the instructor.

2. Never place chemicals directly on the balance pan; first place them on a weighing paper, weighing "boat", or in a container. Clean up any materials you spill on or around the balance.

3. Before moving objects on and off the pan, be sure the balance is in the "arrest" position. When you leave the balance, return the balance to the "arrest" or standby position.

4. Never try to make adjustments on a balance. If it seems out of order, tell your instructor.

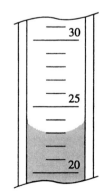

Volume

Beakers and flasks are marked to indicate only approximate volumes. Volume measurements are therefore made in a graduated cylinder by reading the point on the graduated scale that coincides with the bottom of the curved surface called the **meniscus** of the liquid (Figure 2.2). Volumes measured in this illustrated graduated cylinder are calibrated in 1 mL increments and should be estimated and recorded to the nearest 0.1 mL.

Figure 2.2 Read the bottom of the meniscus. The volume is 23.0 mL

Density

Density is a physical property of a substance and is useful in identifying the substance. **Density** is the ratio of the mass of a substance to the volume occupied by that mass; it is the mass per unit volume and is given by the equations

$$\text{Density} = d = \frac{\text{Mass}}{\text{Volume}} = \frac{m}{V} = \frac{g}{mL} \text{ or } \frac{g}{cm^3}$$

In calculating density it is important to make correct use of units and mathematical setups.

Example Problem: An object weighs 283.5 g and occupies a volume of 14.6 mL. What is its density?

$$d = \frac{m}{V} = \frac{283.5 \text{ g}}{14.6 \text{ mL}} = 19.4 \text{ g/mL}$$

Note that all the operations involved in the calculation are properly indicated and that all units are shown. If we divide grams by milliliters, we get an answer in grams per milliliter.

The volume of an irregularly shaped object is usually measured by the displacement of a liquid. An object completely submerged in a liquid displaces a volume of the liquid equal to the volume of the object.

Measurement data and calculations must always be accompanied by appropriate units.

PROCEDURE

Wear protective glasses.

Record your data on the report form as you complete each measurement, never on a scrap of paper which can be lost or misplaced.

A. Temperature

Record all temperatures to the **nearest 0.1°C.**

1. Fill a 400 mL beaker half full of tap water. Place your thermometer in the beaker. Give it a minute to reach thermal equilibrium. Keeping the thermometer in the water and holding the tip of the thermometer away from the glass, read and record the temperature.

2. Fill a 150 mL beaker half full of tap water. Set up a ring stand with the ring and wire gauze at a height so the hottest part of the burner flame will reach the bottom of the beaker. Heat the water to boiling. Read and record the temperature of the boiling water, being sure to hold the thermometer away from the bottom of the beaker.

3. Fill a 250 mL beaker one-fourth full of tap water and add a 100 mL beaker of crushed ice. Without stirring, place the thermometer in the beaker, resting it on the bottom. Wait at least 1 minute, then read and record the temperature. Now stir the mixture for about 1 minute. If almost all the ice melts, add more. Holding the thermometer off the bottom, read and record the temperature. Save the ice-water mixture for Part 4.

4. Weigh approximately 5 g of sodium chloride and add it to the ice-water mixture. Stir for 1 minute, adding more ice if needed. Read and record the temperature. Dispose of the salt water/ice mixture in the sink.

WASTE DISPOSE OF PROPERLY

B. Mass

Using the balance provided, do the following, recording all the masses to include one uncertain digit and all certain digits.

1. Weigh a 250 mL beaker.

2. Weigh a 125 mL Erlenmeyer flask.

3. Weigh a piece of weighing paper or a plastic weighing "boat."

4. Add approximately 2 g of sodium chloride to the weighing paper from step 3 and record the total mass. Calculate the mass of sodium chloride.

C. Length

Using a ruler, make the following measurements in centimeters; measure to the nearest uncertain digit.

1. Measure the length of the arrow on the right   ⟶

2. Measure the external height of a 250 mL beaker.

3. Measure the length of a test tube.

D. Volume

Using the graduated cylinder most appropriate, measure the following volumes to the maximum precision possible, usually 0.1 mL. Remember to read the volume at the meniscus.

1. Fill a test tube to the brim with water and measure the volume of the water.

2. Fill a 125 mL Erlenmeyer flask to the brim with water and measure the volume of the water.

3. Measure 5.0 mL of water in a graduated cylinder and pour it into a test tube. With a ruler, measure the height (in cm) and mark the height with a marker.

4. Measure 10.0 mL of water in the graduated cylinder and pour it into a test tube like the one used in the previous step. Again, mark the height with a marker.

In the future, you will often find it convenient to estimate volumes of 5 and 10 mL simply by observing the height of the liquid in the test tube.

E. Density

Estimate and record all volumes to the highest precision, usually 0.1 mL. Make all weighing to the highest precision of the balance. Note that you must supply the units for the measurements and calculations in this section.

1. Density of Water. Weigh a clean, dry 50 mL graduated cylinder and record its mass. (Graduated cylinders should never be dried over a flame.) Fill the graduated cylinder with distilled water to 50.0 mL. Use a medicine dropper to adjust the meniscus to the 50.0 mL mark. Record the volume. Reweigh and calculate the density of water.

2. Density of a Rubber Stopper. Select a solid rubber stopper which is small enough to fit inside the 50 mL graduated cylinder. Weigh the dry stopper. Fill the 50 mL cylinder with tap water to approximately 25 mL. Read and record the exact volume. Carefully place the rubber stopper into the graduated cylinder so that it is submerged. Read and record the new volume. Calculate the volume and density of the rubber stopper.

3. Density of a Solid Object. Obtain a solid object from your instructor. Record the sample code on the report form. Determine the density of your solid by following the procedure given in Part 2 for the rubber stopper. To avoid the possibility of breakage, incline the graduated cylinder at an angle and slide, rather than drop, the solid into it.

Return the solid object to your instructor.

REPORT FOR EXPERIMENT 2

Measurements

A. Temperature

   1. Water at room temperature   _____ °C

   2. Boiling point   _____ °C

   3. Ice water

      Before stirring   _____ °C

      After stirring for 1 minute   _____ °C

   4. Ice water with salt added   _____ °C

B. Mass

   1. 250 mL beaker   _____ g

   2. 125 mL Erlenmeyer flask   _____ g

   3. Weighing paper or weighing boat   _____ g

   4. Mass of weighing paper/boat + sodium chloride   _____ g

      Mass of sodium chloride (show calculation setup)   _____ g

C. Length

   1. Length of $\longrightarrow$ _____ cm

   2. Height of 250 mL beaker   _____ cm

   3. Length of test tube   _____ cm

D. Volume

   1. Test tube   _____ mL

   2. 125 mL Erlenmeyer flask   _____ mL

   3. Height of 5.0 mL of water in test tube   _____ cm

   4. Height of 10.0 mL of water in test tube   _____ cm

E. Density

1. Density of Water

Mass of empty graduated cylinder   _____

Volume of water   _____

Mass of graduated cylinder and water   _____

Mass of water (show calculation setup)   _____

Density of water (show calculation setup)   _____

2. Density of a Rubber Stopper

Mass of rubber stopper   _____

Initial volume of water in cylinder   _____

Final volume of water in cylinder (including stopper)   _____

Volume of rubber stopper (show calculation setup)   _____

Density of rubber stopper (show calculation setup)   _____

3. Density of a Solid Object

Number of solid object   _____

Mass of solid object   _____

Initial volume of water in graduated cylinder   _____

Final volume in graduated cylinder   _____

Volume of solid object (show calculation setup)   _____

Density of solid object (show calculation setup)   _____

QUESTIONS AND PROBLEMS

1. The directions state "weigh about 5 grams of sodium chloride". Give minimum and maximum amounts of sodium chloride that would satisfy these instructions.

2. Two students each measured the density of a quartz sample three times:

	Student A	Student B	
1.	3.20 g/mL	2.82 g/mL	The density found in the *Handbook*
2.	2.58 g/mL	2.48 g/mL	*of Chemistry and Physics* for quartz
3.	2.10 g/mL	2.59 g/mL	is 2.65 g/mL
mean	2.63 g/mL	2.63 g/mL	

(a) Which student measured density with the greatest precision? Explain your answer.

(b) Which student measured density with the greatest accuracy? Explain your answer.

(c) Are the errors for these students random or systematic? Explain.

Show calculation setups and answers for the following problems.

3. Convert 21°C to degrees Fahrenheit.   _____

4. Convert 101°F to degrees Celsius.   _____

5. An object is 9.6 cm long. What is the length in inches?   _____

6. An empty graduated cylinder weighs 82.450 g. When filled to 50.0 mL with an unknown liquid it weighs 110.810 g. What is the density of the unknown liquid?

_____

7. It is valuable to know that 1 milliliter (mL) equals 1 cubic centimeter (cm^3 or cc). How many cubic centimeters are in an 8.00 oz bottle of cough medicine? (1.00 oz = 29.6 mL)

_____

8. A metal sample weighs 56.8 g. How many ounces does this sample weigh? (1 lb = 16 oz)

_____

9. Convert 15 nm into km.

_____

EXPERIMENT 3

Preparation and Properties of Oxygen

MATERIALS AND EQUIPMENT

Solids: candles, magnesium (Mg) strips, manganese dioxide (MnO_2), fine steel wool (Fe), roll sulfur (S), wood splints. **Solution:** 9 percent hydrogen peroxide (H_2O_2). Deflagration spoon, pneumatic trough, 20 to 25 cm length rubber tubing, 25×200 mm ignition tube, five wide-mouth (gas-collecting) bottles, five glass cover-plates, Büchner funnel, heavy-wall filtering flask with side-arm tubulation, rubber suction tubing, filter paper to fit the Büchner funnel. **Demonstration supplies:** cotton, sodium peroxide (Na_2O_2); steel wool, 25×200 mm test tube; Hoffman electrolysis apparatus.

DISCUSSION

Oxygen is the most abundant and widespread of all the elements in the earth's crust. It occurs both as free oxygen gas and combined in compounds with other elements. Free oxygen gas is diatomic and has the formula O_2. Oxygen is found combined with more elements than any other single element, and it will combine with all the elements except some of the noble gases. Water is 88.9 percent oxygen by mass and the atmosphere is about 21 percent oxygen by volume. Oxygen gas is colorless and odorless, and is only very slightly soluble in water, a property important to its collection in this experiment.

Oxygen may be obtained by decomposing a variety of oxygen-containing compounds. Some of these are mercury(II) oxide (HgO, mercuric oxide), lead(IV) oxide (PbO_2, lead dioxide), potassium chlorate ($KClO_3$), potassium nitrate (KNO_3), hydrogen peroxide (H_2O_2), and water (H_2O).

In this experiment oxygen is produced by decomposing hydrogen peroxide, and five bottles of oxygen will be collected by the downward displacement of water. After collection, some of the physical and chemical properties of oxygen will be observed.

A. Decomposition of Hydrogen Peroxide to Generate Oxygen

Hydrogen peroxide decomposes very slowly at room temperature. The rate of decomposition is greatly increased by adding a catalyst, manganese dioxide. Although manganese dioxide contains oxygen, it is not decomposed under conditions of this experiment. These equations represent the changes that occur.

Word Equation: Hydrogen peroxide $\longrightarrow$ Water + Oxygen

Formula Equation: $2\,H_2O_2(aq) \xrightarrow{\;MnO_2\;} 2\,H_2O(l) + O_2(g)$

B. Collection of Oxygen

The oxygen is collected by a method known as the downward displacement of water. The gas is conducted from a generator to a bottle of water inverted in a pneumatic trough

(Figure 3.1). The oxygen, which is only very slightly soluble in the water, rises in the bottle and pushes the water down and out. Because oxygen is heavier than air, a glass plate is used to cover the opening of the bottle while it is inverted to a right-side-up position and placed on the benchtop until tested.

C. Properties of Oxygen

Like all kinds of matter, oxygen has both physical and chemical properties and you will observe both in this experiment. One outstanding and important chemical property of oxygen is its ability to support combustion. During combustion oxygen is consumed but does not burn and this ability to support combustion is one test for oxygen. Other substances (a wooden splint or a candle, for example) burn in oxygen producing a visible flame and heat. Compounds containing oxygen and one other element are known as oxides. Thus when elements such as sulfur, hydrogen, carbon, and magnesium burn in air or oxygen, they form sulfur dioxide, hydrogen oxide (water), carbon dioxide, and magnesium oxide, respectively. These chemical reactions may be represented by equations; for example:

Word Equation: Sulfur + Oxygen $\longrightarrow$ Sulfur dioxide

Formula Equation: $S(s) + O_2(g) \longrightarrow SO_2(g)$

See Study Aid 2 for a discussion of writing formulas and chemical equations.

PROCEDURE

A. and B. Generation and Collection of Oxygen from Hydrogen Peroxide

Wear protective glasses.
Wash hydrogen peroxide off skin with water immediately.

1. Assemble the apparatus shown in Figure 3.1. It consists of a 250 mL Erlenmeyer flask, two-hole stopper, thistle tube, glass right-angle bend (Figure 1.3B), glass delivery tube with 135 degree bend (Figure 1.3C), and a 20–25 cm length of rubber tubing. The thistle tube should be at least 24 cm (~10 in.) long and be inserted in the rubber stopper so that there is about 3 mm (1/8 in.) clearance between the end of the tube and the bottom of the flask with the stopper in place. Remember to use glycerol when inserting the glass tubing into the rubber stopper and to hold the glass tubing close to the point of insertion.

2. Fill a pneumatic trough with water until the water level is just above the removable shelf. Attach a piece of rubber tubing to the overflow spigot on the trough and put it in the sink so the water will not spill over the edges of the trough onto the counter. Completely fill five wide-mouth bottles with water. Transfer each bottle to the pneumatic trough by covering its mouth with a glass plate, inverting it, and lowering it into the water. Remove the glass plate below the water level. Place two bottles on the shelf in the trough (over the holes), leaving the other three standing for transfer to the shelf when needed.

3. Using a spatula, put a pea-sized quantity of manganese dioxide (MnO_2) in the generator flask. Replace the stopper, stabilize the flask on the ring stand with a clamp, and make sure that all glass-rubber connections are tight. Add 25 mL of water to the flask through the thistle tube. Make sure that the end of the thistle tube is covered with water (to prevent escape of oxygen gas through the thistle tube).

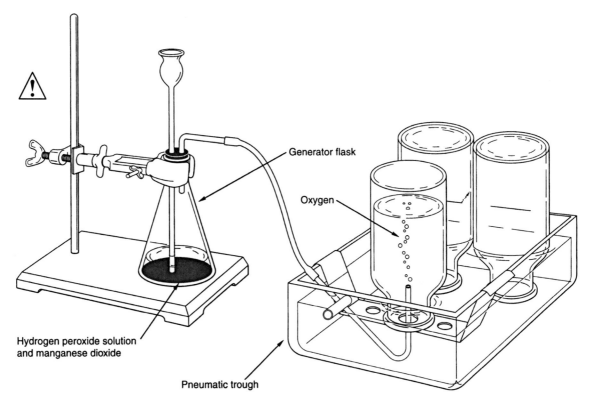

Figure 3.1   Preparing oxygen by decomposing hydrogen peroxide

4. Using a 50 mL graduated cylinder, measure about 50 mL of 9 percent hydrogen peroxide solution.

 Reminder: If hydrogen peroxide gets on your skin, wash it off promptly with water.

To start the generation of oxygen, pour 5 to 10 mL of the peroxide solution into the thistle tube. If all the peroxide solution does not run into the generator, momentarily lift the delivery tube from the water in the trough. Immediately replace the end of the delivery tube under water and into the mouth of the first bottle to collect the gas. When one bottle is filled with gas, immediately start filling the next bottle. Continue generating oxygen by adding an additional 5 to 10 mL portion of hydrogen peroxide whenever the rate of gas production slows down markedly.

5. Cover the mouth of each gas-filled bottle with a glass cover-plate before removing it from the water. Store each bottle mouth upward without removing the glass plate; the oxygen will not readily escape since it is slightly more dense than air. **Note which bottle of gas was collected first** and continue until a total of five bottles of gas have been collected.

6. Allow the reaction to go to completion while you continue with the testing of the oxygen you collected. If you have any unreacted H_2O_2 remaining in the graduated cylinder, return it to the special bottle marked "9% unreacted H_2O_2." When you have completed the rest of this experiment, pour the material in the generator into the vacuum flask through the Büchner funnel setup (see Figure 3.2) for waste MnO_2 disposal. Rinse the generator with water. Occasionally the filter paper will need to be changed and the filter flask emptied into the sink.

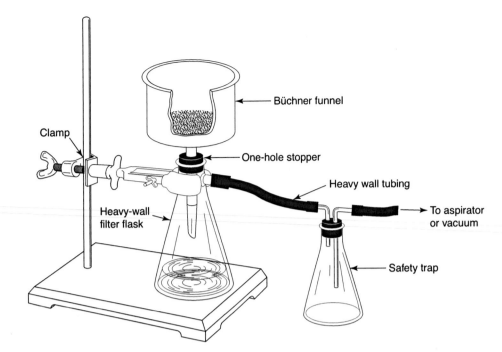

Figure 3.2   Büchner funnel-vacuum flask setup

C.   Properties of Oxygen

Each of the following tests (except C.6) is conducted with a bottle of oxygen and, for comparison, with a bottle of air. Record your observations on the report form.

1. The **glowing splint test** is often used to verify the identity of oxygen. Ignite a wood splint, blow out the flame, and insert the still-glowing splint into the first bottle of oxygen collected. Repeat with a bottle of air. To ensure having a bottle of air, fill the bottle with water and then empty it, thus washing out other gases that may be present.

2. Take a small lump of sulfur in a deflagrating spoon, a bottle of oxygen, and a bottle of air to the fume hood. Light the burner in the fume hood and direct the flame directly into the spoon containing the sulfur. First the sulfur gets dark and melts, then it begins to burn with a blue flame that is barely visible. Lower the burning sulfur alternately into a bottle of oxygen and a bottle of air and compare combustions. Quench the excess burning sulfur in a beaker of water.

3. Stand a small candle (no longer than 5 cm) on a glass plate and light it. Lower a bottle of oxygen over the burning candle, placing the mouth of the bottle on the glass plate. **Measure and record the time,** in seconds, that the candle continues to burn. Repeat with a bottle of air. Note also the difference in the brilliance of the candle flame in oxygen and in air. Return the unused portion of the candle to the reagent shelf.

4. Invert a bottle of oxygen, covered with glass plate, and place it mouth to mouth over a bottle of air. Then remove the glass plate from between the bottles and allow them to stand mouth to mouth for 3 minutes. Cover each bottle with a glass plate and set the bottles down, mouths upward. Test the contents of each bottle by inserting a glowing splint.

5. Pour 25 mL of water into the fifth bottle of oxygen and replace the cover. Place the bottle close to (within 5 or 6 cm) the burner. Take a loose, 4 or 5 cm wad of steel wool (iron) in the crucible tongs and momentarily heat it in the burner flame until some of the steel wool first begins to glow. Immediately lower the glowing metal into the bottle of oxygen. (It is essential that some of the steel wool be glowing when it goes into the oxygen.) Repeat, using a bottle of air.

> **NOTE:** The 25 mL of water is to prevent breakage if the glowing steel wool is accidentally dropped into the bottle.

6. A small strip of magnesium ribbon will be burned next. Read the following precautions before proceeding. *Do not put burning magnesium into a bottle of oxygen.* There is enough oxygen in air for this reaction to proceed vigorously.

 Do not look directly at the burning magnesium ribbon. It is very bright and the light includes considerable ultraviolet light, which can cause damage to the retina of the eye.

Take a 2 to 5 cm strip of magnesium metal in a pair of crucible tongs and ignite it by heating it in the burner flame. After the burning is over, put the product on the Ceramfab plate and compare it to the metal from which it was produced.

D. Instructor Demonstrations (Optional)

1. **Sodium Peroxide as a Source of Oxygen.** Spread some cotton on the bottom of an evaporating dish and sprinkle a small amount (less than 1 g) of fresh sodium peroxide on it. Sprinkle a few drops of water on the peroxide. Spontaneous combustion of the cotton will occur.

2. **Approximate Percentage of Oxygen in the Air.** Push a small wad of steel wool to the bottom of a 25 × 200 mm test tube. Wet the steel wool by covering with water; pour out the surplus water; and place the tube, mouth downward, in a 400 mL beaker half full of water. After the oxygen in the trapped air has reacted with the steel wool—at least three days are needed for complete reaction—adjust the water levels inside and outside the tube to the same height. Cover the mouth of the tube, remove from the beaker, and measure the volume of water in the tube. Alternatively, the height of the water column may be measured (in millimeters) without removing the tube from the beaker. The volume of water in the tube is approximately equal to the volume of oxygen originally present in the tube of air.

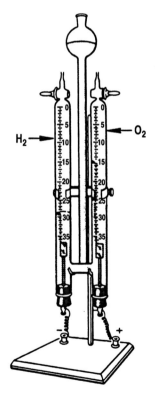

$$\% \text{ oxygen} = \left(\frac{\text{Volume of water in tube}}{\text{Volume of tube}}\right)(100)$$

or

$$\% \text{ oxygen} = \left(\frac{\text{Height of water column}}{\text{Length of tube}}\right)(100)$$

3. **Decomposition of Water.** Set up the Hoffman electrolysis apparatus, as shown in Figure 3.3. The solution used in the apparatus should contain about 2 mL of sulfuric acid per 100 mL of water. Direct current may be obtained from several 1.5 volt type A cells connected in series or from some other D.C. source.

Figure 3.3 Hoffman electrolysis apparatus

REPORT FOR EXPERIMENT 3

Preparation and Properties of Oxygen

A. and B.   Generation and Collection of Oxygen

1. What evidence did you observe that oxygen is not very soluble in water?

2. What is the source of oxygen in the procedure you used?

    Name _____ Formula _____

3. What purpose does the manganese dioxide serve in this preparation of oxygen?

4. What gas was in the apparatus before you started generating oxygen? Where did it go?

5. What is different about the composition of the first bottle of gas collected compared to the other four?

6. Why are the bottles of oxygen stored with the mouth up?

7. (a) What is the symbol of the element oxygen?   _____

    (b) What is the formula for oxygen gas?   _____

8. Which of the following formulas represent oxides? (Circle) MgO, $KClO_3$, SO_2, MnO_2, O_2, $NaOH$, PbO_2, Na_2O_2

9. Write the word and formula equations for the preparation of oxygen from hydrogen peroxide.

Word Equation:

Formula Equation:

10. What substances, other than oxygen, are in the generator when the decomposition of H_2O_2 is complete?

C. Properties of Oxygen

1. Write word equations for the chemical reactions that occurred. (See Study Aid 2.)

C.1. Combustion of wood. Assume carbon is the combustible material.

C.2. Combustion of sulfur.

C.5. Combustion of steel wool (iron). (Call the product iron oxide.)

C.6. Combustion of magnesium.

2. Write formula equations for these four chemical reactions.

C.1. (CO_2 is the formula for the oxide of carbon that is formed.)

C.2. (SO_2 is the formula for the oxide of sulfur that is formed.)

C.5. (Fe_3O_4 is the formula for the oxide of iron that is formed.)

C.6. (MgO is the formula for the oxide of magnesium that is formed.)

3. Combustion of a candle.

   (a) Number of seconds that the candle burned in the bottle of oxygen.   _____

   (b) Number of seconds that the candle burned in the bottle of air.   _____

   (c) Explain this difference in combustion time.

   (d) Is it scientifically sound to conclude that all the oxygen in the bottle was reacted when the candle stopped burning? Explain.

4. What were the results of the experiment in which a bottle of oxygen was placed over a bottle of air? Explain the results.

5. (a) Describe the material that is formed when magnesium is burned in air.

   (b) What elements are in this product?

6. (a) What is your conclusion about the rate or speed of a chemical reaction with respect to the concentration of the reactants—for example, a combustion in a high concentration of oxygen (pure oxygen) compared to a combustion in a low concentration of oxygen (air)?

(b) What evidence did you observe in the burning of sulfur to confirm your conclusion in 6(a)?

EXPERIMENT 6

Freezing Points — Graphing of Data

MATERIALS AND EQUIPMENT

Solids: benzoic acid (C_6H_5COOH) and crushed ice. **Liquid:** glacial acetic acid ($HC_2H_3O_2$). Thermometer, watch or clock with second hand, slotted corks or stoppers.

DISCUSSION

All pure substances, elements and compounds, possess unique physical and chemical properties. Just as one human being can be distinguished from all others by certain characteristics — fingerprints or DNA, for example — it is also possible, through knowledge of its properties, to distinguish any given compound from among the many millions that are known.

A. Melting and Freezing Points of Pure Substances

The melting point and the boiling point are easily determined physical properties that are very useful in identifying a substance. Consequently, these properties are almost always recorded when a compound is described in the chemical literature (textbooks, handbooks, journal articles, etc.). The freezing and melting of a pure substance occurs at the same temperature, measured when the liquid and solid phases of the substance are in equilibrium. When energy is being removed from a liquid in equilibrium with its solid, the process is called freezing; when energy is being added to a solid in equilibrium with its liquid, the process is called melting.

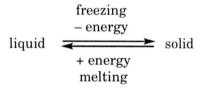

In this experiment, we will determine the freezing point of a pure organic compound, glacial acetic acid ($HC_2H_3O_2$). When the experimental freezing point has been determined, it will be compared with the melting point temperature listed in the *Handbook of Chemistry and Physics.*

When heat is removed from a liquid, the liquid particles lose kinetic energy and move more slowly causing the temperature of the liquid to decrease. Finally enough heat is removed and the particles move so slowly that the liquid becomes a solid, often a crystalline solid. The temperature when this happens (the freezing point) is different for different substances.

The amount of energy removed from a quantity of liquid to freeze it, is equal to the amount of energy added to the same quantity of its solid to melt it. Thus, depending on the direction of energy flow, this equilibrium temperature is called the melting point or the freezing point.

B. Freezing Point of Impure Substances

When a substance (solvent) is uniformly mixed with a small amount of another substance (solute), the freezing point of the resulting solution (an "impure substance") will be lower than that of the pure solvent. For example, the accepted freezing point for pure water is 0.0°C. Solutions of salt in water may freeze at temperatures as low as −21°C depending on the amount of salt added to the water. Antifreeze is added to the water in a car radiator to lower the freezing point of the water.

Melting point/freezing point data are of great value in determining the identity and/or purity of substances, especially in the field of organic chemistry. If a sample of a compound melts or freezes appreciably below the known melting point of the pure substance, we know that the sample contains impurities which have lowered the melting point. If the melting point of an unknown compound agrees with that of a known compound, the identity can often be confirmed by mixing the unknown compound with the known and determining the melting point of the mixture. If the melting point of the mixture is the same as that of the known compound, the compounds are identical. On the other hand, a lower melting point for the mixture indicates that the two compounds are not identical.

C. Supercooling During Freezing

Frequently when a substance is being cooled, the temperature will fall below the true freezing point before crystals begin to form. This phenomenon is known as supercooling because the substance is cooled below its freezing point without forming a solid. Supercooling is more likely to occur if the liquid remains very still and undisturbed as its temperature is lowered. When the system is disturbed in any way, for example, by stirring or jarring, crystallization occurs rapidly throughout the system. As the crystals form, heat is released (called the heat of crystallization) and the temperature rises quickly to the freezing point of the substance. Thus, supercooling does not change the freezing point of the substance.

D. Freezing Point Determinations

You will do three freezing point determinations during this experiment using the setup in Figure 6.1.

Trial 1. Freezing point determination of pure glacial acetic acid WITH STIRRING. This will usually eliminate supercooling.

Trial 2. Freezing point determination of pure glacial acetic acid WITHOUT STIRRING. This should enhance the possibility of supercooling but does not guarantee it.

Trial 3. Freezing point determination of acetic acid (the solvent) after benzoic acid (a solute) has been dissolved in it. This will be done WITHOUT STIRRING to enhance supercooling again.

The time/temperature data will be graphed and the freezing point for each trial read from the graph.

PROCEDURE

Wear protective glasses.

> **NOTES:** Since water and other contaminants will influence the freezing points in this experiment, use only clean, dry equipment.
>
> Read and record all temperatures to the nearest 0.1°C.

A. Freezing Point Determination of Pure Glacial Acetic Acid

Trial 1: With stirring

1. Fasten a utility clamp to the top of a clean, dry test tube. Position this clamp-tube assembly on a ring stand so that the bottom of the tube is about 20 cm above the ring stand base.

2. Obtain a slotted one-hole cork (or stopper) to fit the test tube (see Figure 6.1). Insert a thermometer in the cork and position it in the test tube so that the end of the bulb is about 1.5 cm from the bottom of the test tube. Turn the thermometer so that the temperature scale can be read in the slot.

3. Take your test tube, the cork/thermometer and a graduated cylinder to the fume hood. Measure out 10. mL of glacial acetic acid. Pour it into the test tube and close the test tube with the cork/thermometer. Glacial acetic acid is irritating and harmful if inhaled so keep the test tube stoppered while you work outside the hood at your bench. Rinse the graduated cylinder with water immediately.

4. Reclamp the test tube to your ring stand to minimize the risk of spilling. Make sure the thermometer bulb is covered by the acid and adjust the temperature of the acetic acid to approximately 25°C by warming or cooling the tube in a beaker of water.

5. Fill a 400 mL beaker about three-quarters full of crushed ice; add cold water until the ice is almost covered. Position the beaker of ice and water on the ring stand base under the clamped tube-thermometer assembly.

6. Read the temperature of the acetic acid and record as the 0.0 minute time reading in the Data Table. Now loosen the clamp on the ring stand and observe the second hand of your watch or clock. As the second hand crosses 12, lower the clamped tube-thermometer assembly so that all of the acetic acid in the tube is below the surface of the ice water. Fasten the clamp to hold the tube in this position.

7. Loosen the cork on the tube and stir (during Trial 1 only) the acid with the thermometer, keeping the bulb of the thermometer completely immersed in the acid. Take accurate temperature readings at 30-second intervals as the acid cools. (Zero time was when the second hand crossed 12.) Stop stirring and center the thermometer bulb in the tube as soon as you are sure that crystals are forming in the acid (one to four minutes). Circle the temperature reading when the crystals were first observed.

8. Continue to take temperature readings at 30-second intervals until a total time of 12 minutes has elapsed or until the entire volume of liquid becomes solidified. After that occurs, read the temperature for an additional 2 minutes (4 time intervals) and continue with the next step.

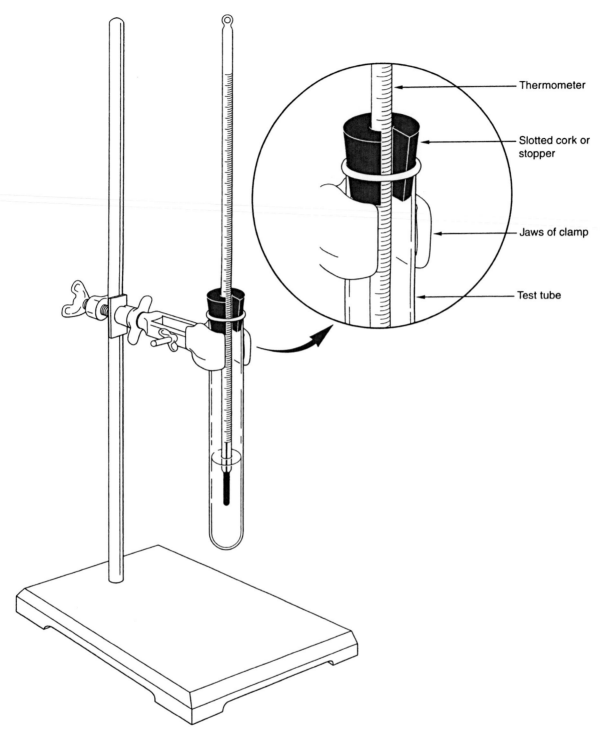

Figure 6.1   Setup for freezing-point determination

9. After completing the temperature readings, remove the test tube-thermometer assembly from the ice bath, keeping the thermometer in place. Immerse the lower portion of the test tube in a beaker of warm water to melt the frozen acetic acid. Do not discard this acid; it will be used in Trials 2 and 3.

Trial 2: Without stirring

10. Repeat steps 4–9 with the following changes:

   a. Replenish the ice bath as in step 5.

   b. After submerging the tube in the ice bath, do NOT stir. Do NOT touch or move the apparatus in any way

   c. If the temperature goes down to about 4°C or lower without the formation of acetic acid crystals and remains there, touch the thermometer and move it until crystals form which usually happens quickly. When you do this be very observant of the temperature changes. Continue to record temperature readings for the full 12 minutes or until the temperature stabilizes after crystallization for 5 minutes.

B. Freezing Point Determination of An Acetic Acid/Benzoic Acid Solution

Trial 3: Without stirring

11. Weigh approximately 0.50 g (between 0.48 and 0.52 g) of benzoic acid crystals. Now remove the thermometer from the test tube of acetic acid and lay it on the table) being careful not to contaminate the thermometer or lose any acid. Carefully add all of the benzoic acid to the acetic acid. Stir gently with the thermometer until all of the crystals have dissolved. Stir for an additional minute or two to ensure a uniform solution. Adjust the temperature of the solution to approximately 25°C.

12. Repeat step 10.

13. Warm the test tube to change the solid to a liquid and dispose of the acetic acid/benzoic acid solution in the waste container provided. Rinse the test tube with water and pour the liquid down the sink.

C. Graphing Temperature Data

Graph the three sets of data using the graph paper in the report form or prepare a computer graph. If necessary, review the instructions for preparing a graph in Study Aid 3.

NAME _____

SECTION _____ DATE _____

INSTRUCTOR _____

REPORT FOR EXPERIMENT 6

Freezing Points–Graphing of Data

Data Table

time, minutes	Pure Acetic Acid temp, °C WITH STIRRING	Pure Acetic Acid temp, °C WITHOUT STIRRING	Impure Acetic Acid temp, °C WITHOUT STIRRING
0.0			
0.5			
1.0			
1.5			
2.0			
2.5			
3.0			
3.5			
4.0			
4.5			
5.0			
5.5			
6.0			
6.5			
7.0			
7.5			
8.0			
8.5			
9.0			
9.5			
10.0			
10.5			
11.0			
11.5			
12.0			

Graphing of Freezing Point Data

Plot your data on the graph paper or the computer using a legend as follows:

△ = Pure acetic acid with stirring

▲ = Pure acetic acid without stirring

○ = Acetic acid/benzoic acid solution without stirring

Draw rectangles around the portions of your curves that show supercooling.

QUESTIONS

Use your graph to answer the questions 1–3.

1. a. At what temperature did crystals first form in Trial 1?   _____

   b. Where did the temperature stabilize after supercooling in Trial 2?   _____

   c. What is your experimental freezing point of glacial acetic acid?   _____

   d. What is the theoretical freezing point of glacial acetic acid?   _____
      (Consult the *Handbook of Chemistry and Physics*)

2. How many degrees was the freezing point depressed by the benzoic acid?   _____

   Do this by estimating to the nearest 0.1 degree the number of degrees between the flattest (most nearly horizontal) portions of the curves. Mark the area on the graph with an arrow (↓) to show where this temperature difference estimate was made.

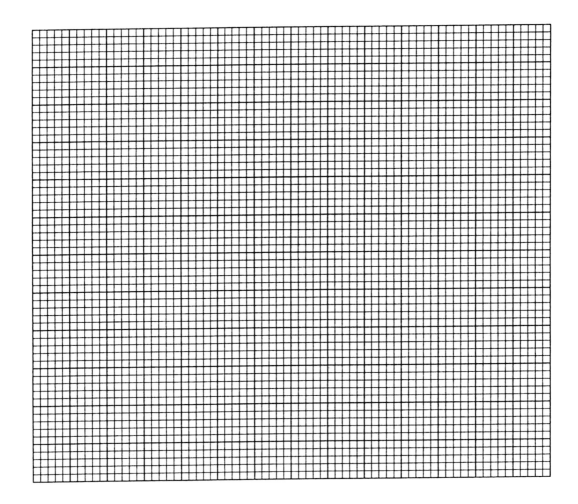

3. a. What is the effect of stirring on the freezing point of pure acetic acid?

   b. What is the effect of stirring on supercooling?

4. a. What do the melting point and freezing point of a substance have in common?

   b. What is the difference between the melting and freezing of a substance?

5. When the solid and liquid phases are in equilibrium, which phase, solid or liquid contains the greater amount of energy? Explain the rationale for your answer.

EXPERIMENT 7

Water in Hydrates

MATERIALS AND EQUIPMENT

Solids: finely ground copper(II) sulfate pentahydrate ($CuSO_4 \cdot 5H_2O$), and unknown hydrate. Cobalt chloride test paper, clay triangle, crucible and cover, 25×200 mm ignition test tube, watch glass.

DISCUSSION

Many salts form compounds in which a definite number of moles of water are combined with each mole of the anhydrous salt. Such compounds are called **hydrates.** The water which is chemically combined in a hydrate is referred to as **water of crystallization** or **water of hydration.** The following are representative examples:

$$CaSO_4 \cdot 2H_2O, \quad CoCl_2 \cdot 6H_2O, \quad MgSO_4 \cdot 7H_2O, \quad Na_2CO_3 \cdot 10H_2O$$

In a hydrate the water molecules are distinct parts of the compound but are joined to it by bonds that are weaker than either those forming the anhydrous salt or those forming the water molecules. In the formula of a hydrate a dot is commonly used to separate the formula of the anhydrous salt from the number of molecules of water of crystallization. For example, the formula of calcium sulfate dihydrate is written $CaSO_4 \cdot 2H_2O$ rather than $CaSO_6H_4$.

Hydrated salts can usually be converted to the anhydrous form by careful heating:

$$\text{Hydrated Salt} \xrightarrow{\Delta} \text{Anhydrous salt} + \text{water}$$

Hydrated salts can be studied qualitatively and quantitatively. In the **qualitative** part of this experiment we will observe some of properties of the liquid (water) driven off by heating the sample. In the **quantitative** part of the experiment we will determine **how much** water was in the hydrate by measuring the amount of water driven off by heating.

To make certain that all of the water in the original sample has been driven off, chemists use a technique known as **heating to constant weight.** Since time expended for this is limited, constant weight is essentially achieved when the sample is heated and weighed in successive heatings until the weight differs by no more than 0.05 g. Thus, if the second weighing is no more than 0.05 g less than the first heating, a third heating is not necessary because the sample has been heated to constant weight (almost). This is a very good reason to follow directions meticulously when heating. If the sample is not heated long enough or at the correct temperature, all of the water may not be driven off completely in the first heating.

Hence it is possible to determine the percentage of water in a hydrated salt by determining the amount of mass lost (water driven off) when a known mass of the hydrate is heated to constant weight.

$$\text{Percentage water} = \left(\frac{\text{Mass lost}}{\text{Mass of sample}} \right)(100)$$

It is possible to condense the vapor driven off the hydrate and demonstrate that it is water by testing it with anhydrous cobalt(II) chloride ($CoCl_2$). Anhydrous cobalt(II) chloride is blue but reacts with water to form the red hexahydrate, $CoCl_2 \cdot 6H_2O$.

PROCEDURE

Wear protective glasses.

A. Qualitative Determination of Water

1. Fold a 2.5 × 20 cm strip of paper lengthwise to form a V-shaped trough or chute. Load about 4 g of finely ground copper(II) sulfate pentahydrate in this trough, spreading it evenly along the length of the trough.

2. Clamp a **dry** 25 × 200 mm ignition test tube so that its mouth is 15–20 degrees **above the horizontal** (Figure 7.1a). Insert the loaded trough into the tube. Rotate the tube to a nearly vertical position (Figure 7.1b) to deposit the copper(II) sulfate in the bottom of the tube. Tap the paper chute gently if necessary, but make sure that no copper sulfate is spilled and adhering to the sides of the upper part of the tube.

3. Remove the chute and turn the tube until it slants mouth downward at an angle of 15–20 degrees **below the horizontal** (Figure 7.1c). Make sure that all of the copper(II) sulfate remains at the bottom of the tube. To obtain a sample of the liquid that will condense in the cooler part of the tube, place a clean, dry test tube, held in an upright position in either a rack or an Erlenmeyer flask, just below the mouth of the tube containing the hydrate.

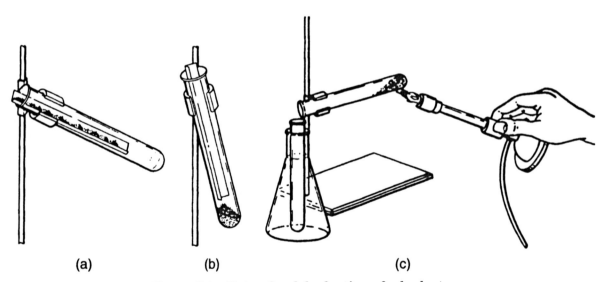

(a) (b) (c)

Figure 7.1 Setup for dehydration of a hydrate

4. Heat the hydrate gently at first to avoid excessive spattering. Gradually increase the rate of heating, noting any changes that occur and collecting some of the liquid that condenses in the cooler part of the tube. Continue heating until the blue color of the hydrate has disappeared, but do not heat until the residue in the tube has turned black. Finally warm the tube over its entire length—without directly applying the flame to the clamp—for a minute

or two to drive off most of the liquid that has condensed on the inner wall of the tube. Allow the tube and contents to cool.

NOTE: At excessively high temperatures (above 600°C) copper(II) sulfate decomposes; sulfur trioxide is driven off and the black copper(II) oxide remains as a residue.

Observe and record the appearance and odor of the liquid that has been collected.

5. While the tube is cooling, dry a piece of cobalt chloride test paper by holding it with tongs about 20 to 25 cm above a burner flame; that is, close enough to heat but not close enough to char or ignite the paper. When properly dried, the test paper should be blue. Using a clean stirring rod, place a drop of the liquid collected from the hydrate on the dried cobalt chloride test paper. For comparison place a drop of distilled water on the cobalt chloride paper. Record your observations.

6. Empty the anhydrous salt residue in the tube onto a watch glass and divide it into two portions. Add 3 or 4 drops of the liquid collected from the hydrate to one portion and 3 or 4 drops of distilled water to the other. Compare and record the results of these tests.

 Dispose of solid residues in the waste heavy metal container provided.

B. Quantitative Determination of Water in a Hydrate

NOTES:

1. **Weigh crucible and contents to the highest precision with the balance available to you.**

2. Since there is some inaccuracy in any balance, use the same balance for successive weighings of the same sample. When subtractions are made to give mass of sample and mass lost, the inaccuracy due to the balance should cancel out.

3. Handle crucibles and covers with tongs only, after initial heating.

4. Be sure crucibles are at or near room temperature when weighed.

5. **Record all data directly on the report form as soon as you obtain them.**

---

1. Obtain a sample of an unknown hydrate, as directed by your instructor. Be sure to record the identifying number.

2. Weigh a clean, dry crucible and cover to the highest precision of the balance.

3. Place between 2 and 3 g of the unknown into the weighed crucible. Cover and weigh the crucible and contents.

4. Place the covered crucible on a clay triangle; adjust the cover so that is slightly ajar, to allow the water vapor to escape (see Figure 7.2); and **very gently** heat the crucible for about 5 minutes. Readjust the flame so that a sharp, inner-blue cone is formed. Heat for another 12 minutes with the tip of the inner-blue cone just touching the bottom of the crucible. The crucible bottom should become dull red during this period.

5. After this first heating is completed, close the cover, cool (about 10 minutes), and weigh.

6. To determine if all the water in the sample was removed during the initial heating, reheat the covered crucible and contents for an additional 6 minutes at maximum temperature; cool and reweigh. If the sample was heated to constant weight the results of the last two weighings should agree within 0.05 g. If the decrease in mass between the two weighings is greater than 0.05 g, repeat the heating and weighing until the results of two successive weighings agree to within 0.05 g.

7. Calculate the percentage of water in your sample on the basis of the *final* weighing.

 Dispose of the solid residue in the waste heavy metal container provided. Return the unused portion of your unknown to the instructor.

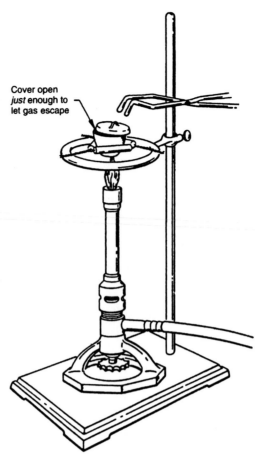

Cover open *just* enough to let gas escape

Figure 7.2   Method of heating a crucible

EXPERIMENT 9

Properties of Solutions

MATERIALS AND EQUIPMENT

Solids: ammonium chloride (NH_4Cl), barium chloride ($BaCl_2$), barium sulfate ($BaSO_4$), fine and coarse crystals of sodium chloride ($NaCl$), and sodium sulfate (Na_2SO_4). **Liquids:** decane ($C_{10}H_{22}$), isopropyl alcohol (C_3H_7OH), and kerosene. **Solutions:** saturated iodine-water (I_2), and saturated potassium chloride (KCl).

DISCUSSION

Solute, Solvent, and Solution

The term **solution** is used in chemistry to describe a homogeneous mixture in which at least one substance (the **solute**) is dissolved in another substance (the **solvent**). The solvent is the substance present in greater quantity and the name of the solution is taken from the name of the solute. Thus, when sodium chloride is dissolved in water, sodium chloride is the solute, water is the solvent, and the solution is called a sodium chloride solution.

In this experiment we will be working with two common types of solutions: those in which a solid solute is dissolved in a liquid solvent (water), and a few in which a liquid solute is dissolved in a liquid solvent.

Like other mixtures, a solution has variable composition, since more or less solute can be dissolved in a given quantity of a solvent. The amount of solute that remains uniformly dispersed throughout the solution after mixing is referred to as the **solution concentration** and can be expressed in many different ways. The maximum concentration that a solution can have varies depending on many factors, including the temperature, the kind of particles in the solute, and interactions between the solute particles and the solvent. In general, water, which is polar, is a better solvent for inorganic than for organic substances. On the other hand, nonpolar solvents such as benzene, decane, and ether are good solvents for many organic substances that are practically insoluble in water.

Dissolved solute particles can be either molecules or ions and their size is of the order of 10^{-8} to 10^{-7} cm (1-10 Å). Many substances will react chemically with each other only when they are dissociated into ions in solution. For example, when the two solids sodium chloride ($NaCl$) and silver nitrate ($AgNO_3$) are mixed, no detectable reaction is observed. However, when aqueous solutions of these salts are mixed, their component ions react immediately to form a white precipitate ($AgCl$).

The rate at which a solute and solvent will form a solution depends on several factors, all of which are related to the amount of contact between the solute particles and the solvent. A solid can dissolve only at the surface that is in contact with the solvent. Any change which increases that contact will increase the rate of solution and vice versa. Thus, the rate of dissolving a solid solute depends on:

1. The particle size of the solute
2. Agitation or stirring of the solution
3. The temperature of the solution
4. The concentration of the solute in solution

Solubility, Miscibility, and Concentration

The term **solubility** refers to the maximum amount of solute that will dissolve in a specified amount of solvent under stated conditions. At a specific temperature, there is a limit to the amount of solute that will dissolve in a given amount of solvent.

Solubility can be expressed in a relative, qualitative way. For example a solute may be very soluble, moderately soluble, slightly soluble, or insoluble in a given solvent at a given temperature. Table 8.1 shows how temperature effects the amount of four different salts that dissolve in 100 g of water.

Table 9.1
Temperature Effect on Solubility of Four Salts in Water, g solute/100 g water

	0°C	10°C	20°C	30°C	40°C	50°C	60°C	70°C	80°C	90°C	100°C
KCl	27.6	31.0	34.0	37.0	40.0	42.6	45.5	48.3	51.1	54.0	55.6
NaCl	35.7	35.8	36.0	36.3	36.6	37.0	37.3	37.8	38.4	39.0	39.8
KBr	53.5	59.5	65.2	70.6	75.5	80.2	85.5	90.0	95.0	99.2	104.0
BaCl$_2$	31.6	33.3	35.7	38.2	40.7	43.6	46.6	49.4	52.6	55.7	58.8

The term **miscibility** describes the solubility of two liquids in each other. When both the solute and solvent are liquids, their solubility in each other is described as miscible (soluble) or immiscible (insoluble). For example, ethyl alcohol and water are miscible; oil and water are immiscible.

The **concentration** of a solution expresses how much solute is dissolved in solution and can be expressed several ways:

1. **Dilute vs. Concentrated:** a dilute solution contains a relatively small amount of solute in a given volume of solution; a concentrated solution contains a relatively large amount of solute per unit volume of solution.

2. **Saturated vs. Unsaturated vs. Supersaturated:**

a. A **saturated** solution contains as much dissolved solute as possible at a given temperature and pressure. The dissolved solute is in equilibrium with undissolved solute. A saturated solution can be dilute or concentrated. The solutions described in Table 8.1 are saturated at each temperature.

Solute(solid) $\rightleftharpoons$ Solute(dissolved)

b. **Unsaturated** solutions contain less solute per unit volume than the corresponding saturated solution. Thus, more solute will dissolve in an unsaturated solution (until saturation is reached).

c. **Supersaturated** solutions contain more dissolved solute than is normally present in the corresponding saturated solution. However, a supersaturated solution is in a very unstable state and will form a saturated solution if disturbed. For example, when a small crystal of the dissolved salt is dropped into a supersaturated solution, crystallization begins at once and salt precipitates until a saturated solution is formed.

3. **Mass-percent Solution** is a quantitative expression of concentration expressed as the percent by mass of the solute in a solution. For example, a 10% sodium hydroxide solution contains 10 g of NaOH in 100 g of solution (10 g NaOH + 90 g H_2O); 2 g NaOH in 20 g of solution (2 g NaOH + 18 g H_2O). The formula for calculating mass percent is:

$$\text{Mass percent} = \left(\frac{\text{g solute}}{\text{g solute} + \text{g solvent}}\right)(100)$$

4. **Mass per 100 g solvent** is another quantitative expression of concentration (and the one used in Table 8.1). It is not the same as the Mass percent concentration above because the units are g solute/100 g solvent. Thus, for the 10% NaOH solution described in No. 3, the g NaOH/100 g H_2O would be calculated as follows:

$$\left(\frac{10\,\text{g NaOH}}{90\,\text{g H}_2\text{O}}\right)(100) = \frac{11\,\text{g NaOH}}{100\,\text{g H}_2\text{O}}$$

5. **Molarity** is the most common quantitative expression of concentration. Molarity is the number of moles (molar mass) of solute per liter of solution. Thus a solution containing 1 mole of NaOH (40.00 g) per liter is 1 molar (abbreviated 1 M). The concentration of a solution containing 0.5 mole in 500 mL (0.5 L) is also 1 M. The formula for calculating molarity is:

$$\text{Molarity} = \frac{\text{moles of solute}}{\text{liter of solution}} = \frac{\text{moles}}{\text{liter}}$$

PROCEDURE

Wear protective glasses.

A. Concentration of a Saturated Solution

Use the same balance for all weighings.
Make all weighings to the highest precision of the balance.

1. Prepare a water bath with a 400 mL beaker half full of tap water and heat to boiling. (See Figure 1.6.)

2. Weigh an empty evaporating dish. Obtain 6 mL of saturated potassium chloride solution and pour it into the dish. Weigh the dish with the solution in it and record these masses on the report form.

3. Place the evaporating dish on the beaker of boiling water and continue to boil until the potassium chloride solution has evaporated almost to dryness (about 25 to 30 minutes), **adding more water to the beaker as needed.**

While the evaporation is proceeding, continue with other parts of the experiment.

---

4. Remove the evaporating dish and beaker from the wire gauze and dry the bottom of the dish with a towel. Put the dish on the wire gauze and heat gently for 1-2 minutes to evaporate the last traces of water. Do not heat too strongly because at high temperatures there is danger of sample loss by spattering.

5. Allow the dish with dry potassium chloride to cool on the Ceramfab pad for 5 to 10 minutes and weigh. To be sure that all the water has evaporated from the potassium chloride, put the dish back on the wire gauze and heat gently again for 1-2 minutes.

6. Allow the dish to cool again on the Ceramfab pad for 5 to 10 minutes and reweigh. The second weighing should be no more than 0.05 g less than the first weighing. Otherwise a third heating and weighng should be done.

 7. Add water to the residue in the dish to redissolve the potassium chloride. Pour the solution into the sink and flush generously with water.

B.   Relative Solubility of a Solute in Two Solvents

1. Add about 2 mL of decane and 5 mL of water to a test tube, stopper it, and shake gently for about 5 seconds. Allow the liquid layers to separate and note which liquid has the greatest density.

2. Now, add 5 mL of saturated iodine-water to the test tube, note the color of each layer, insert the stopper, and shake gently for about 20 seconds. Allow the liquids to separate and again note the color of each layer.

 3. Dispose of the mixture in this test tube in the bottle labeled **Decane Waste.**

C.   Miscibility of Liquids

1. Take three dry test tubes and add liquids to each as follows:

    a. 1 mL kerosene and 1 mL isopropyl alcohol

    b. 1 mL kerosene and 1 mL water

    c. 1 mL water and 1 mL isopropyl alcohol

2. Stopper each tube and mix by shaking for about 5 seconds. Note which pairs are miscible. Dispose of the kerosene mixtures (a and b) in the bottle labeled **Kerosene Waste.** Dispose the contents in test tube (c) in the sink.

D. Effect of Particle Size on Rate of Dissolving

1. Fill a dry test tube to a depth of about 0.5 cm with fine crystals of sodium chloride. Fill another dry tube to the same depth with coarse sodium chloride crystals. Add 10 mL of tap water to each tube and stopper. Shake both tubes at the same time, noting the number of seconds required to dissolve the salt in each tube. (Don't shake the tubes for more than two minutes.)

 2. Dispose of these solutions in the sink.

E. Effect of Temperature on Rate of Dissolving

1. Weigh two 0.5 g samples of fine sodium chloride crystals.

2. Take a 100 mL and a 150 mL beaker and add 50 mL tap water to each. Heat the water in the 150 mL beaker to boiling and allow it to cool for about 1 minute.

3. Add the 0.5 g samples of salt to each beaker and observe the time necessary for the crystals to dissolve in the hot water (do not stir).

4. As soon as the crystals are dissolved in the hot water, take the beaker containing the hot solution in your hand, slowly tilt it back and forth, and observe the layer of denser salt solution on the bottom. Repeat with the cold-water solution.

 5. Dispose of these solutions in the sink.

F. Solubility versus Temperature; Saturated and Unsaturated Solutions

1. Label four weighing boats or papers as follows and weigh the stated amounts onto each one.

   a. 1.0 g NaCl   b. 1.4 g NaCl   c. 1.0 g NH_4Cl d. 1.4 g NH_4Cl

2. Record observations in the table provided on the report form as you proceed through 3-6.

3. Add the 1.0 g samples of NaCl and NH_4Cl to separate tubes labeled A and B as shown. Add 5 mL of distilled water to each, stopper and shake until each salt is dissolved.

4. Now add 1.4 g NaCl to test tube A. Add 1.4 g NH_4Cl to test tube B. Stopper and shake for about 3 minutes. Note whether all of the crystals have dissolved.

5. Place both tubes (unstoppered) into a beaker of boiling water, shake occasionally, and note the results after about 5 minutes.

6. Remove the tubes and cool in running tap water for about 1 minute. Let stand for a few minutes and record what you observe.

 7. Dispose of these solutions in the sink. Flush generously with water.

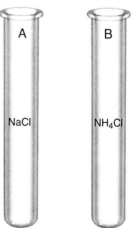

G. Ionic Reactions in Solution

1. Into four labeled test tubes, place pea-sized quantities of the following salts, one salt in each tube: (a) barium chloride, (b) sodium sulfate, (c) sodium chloride, (d) barium sulfate.

2. Add 5 mL of water to each tube, stopper, and shake to dissolve. One of the four salts does not dissolve.

3. Mix the barium chloride and sodium sulfate solutions together. Note the results. (Sodium chloride and barium sulfate are the products of this reaction.)

 Dispose of all tubes containing barium in the waste bottle provided. The remaining tubes can be rinsed in the sink.

REPORT FOR EXPERIMENT 9

Properties of Solutions

A. Concentration of Saturated Solution

1. Mass of empty evaporating dish   _____

2. Mass of dish + saturated potassium chloride solution   _____

3. Mass of dish + dry potassium chloride, 1st heating   _____

4. Mass of dish + dry potassium chloride, 2nd heating   _____

5. Mass of saturated potassium chloride solution   _____
   Show Calculation Setup

6. Mass of potassium chloride in the saturated solution   _____
   Show Calculation Setup

7. Mass of water in the saturated potassium chloride solution   _____
   Show Calculation Setup

8. Mass percent of potassium chloride in the saturated solution   _____
   Show Calculation Setup

9. Grams of potassium chloride per 100 g of water (experimental)   _____
   in the original solution.
   Show Calculation Setup

10. Grams of potassium chloride per 100 g of water (theoretical)   _____
    (From Table 8.1) at 20°C.

B. Relative Solubility of a Solute in Two Solvents

1. (a) Which liquid is denser, decane or water?   _____

(b) What experimental evidence supports your answer?

2. Color of iodine in water: _____

Color of iodine in decane: _____

3. (a) In which of the two solvents used is iodine more soluble?   _____

(b) Cite experimental evidence for your answer.

C. Miscibility of Liquids

1. Which liquid pairs tested are miscible?

2. How do you classify the liquid pair decane—H_2O, miscible or immiscible?

_____

D. Rate of Dissolving Versus Particle Size

1. Time required for fine salt crystals to dissolve   _____

2. Time required for coarse salt crystals to dissolve   _____

3. Since the amount of salt, the volume of water, and the temperature of the systems were identical in both test tubes, how do you explain the difference in time for dissolving the fine vs. the coarse salt crystals?

E. Rate of Dissolving Versus Temperature

1. Under which condition, hot or cold, did the salt dissolve faster?   _____

2. Since the amount of salt, the volume of water, and the texture of the salt crystals were identical in both best tubes, how do you explain the difference in time for dissolving at the hot vs. cold temperatures?

F. Solubility vs. Temperature; Saturated and Unsaturated Solutions

Data Table: Circle the choices which best describe your observations.

	NaCl	NH$_4$Cl
1.0 g + 5 mL water	dissolved completely? yes/no saturated or unsaturated?	dissolved completely? yes/no saturated or unsaturated?
1.0 g + 5 mL water + 1.4 g	dissolved completely? yes/no saturated or unsaturated?	dissolved completely? yes/no saturated or unsaturated?
2.4 g + 5 mL water + heat	dissolved completely? yes/no saturated or unsaturated?	dissolved completely? yes/no saturated or unsaturated?
2.4 g + 5 mL water after cooling	dissolved completely? yes/no saturated or unsaturated?	dissolved completely? yes/no saturated or unsaturated?

G. Ionic Reactions in Solution

1. Write the word and formula equations representing the chemical reaction that occurred between the barium chloride solution, BaCl$_2$(aq), and the sodium sulfate solution, Na$_2$SO$_4$(aq).

   Word Equation:

   Formula Equation:

2. (a) Which of the products is the white precipitate?   _____

   (b) What experimental evidence leads you to this conclusion?

SUPPLEMENTARY QUESTIONS AND PROBLEMS

1. Use the solubility data in Table 9.1 to answer the following:
   Show Calculations

   (a) What is the percentage by mass of NaCl in a saturated solution of sodium chloride at 50°C?

                                                                    _____

(b) Calculate the solubility of potassium bromide at 23°C. Hint: Assume that the solubility increases by an equal amount for each degree between 20°C and 30°C.

_____

(c) A saturated solution of barium chloride at 30°C contains 150 g water. How much additional barium chloride can be dissolved by heating this solution to 60°C?

_____

2. A solution of KCl is saturated at 50°C.
   Use Table 9.1

   (a) How many grams of solute are dissolved in 100 g of water?   _____

   (b) What is the total mass of the solution?   _____

   (c) What is the mass percent of this solution at 50°C?   _____

   (d) If the solution is heated to 100°C, how much more KCl can be dissolved in the solution without adding more water?

   _____

   (e) If the solution is saturated at 100°C and then cooled to 30°C, how many grams of solute will precipitate out?

   _____

EXPERIMENT 10

Composition of Potassium Chlorate

MATERIALS AND EQUIPMENT

Solids: Reagent Grade potassium chlorate ($KClO_3$) and potassium chloride (KCl). **Solutions:** dilute (6 M) nitric acid (HNO_3) and 0.1 M silver nitrate ($AgNO_3$). Two No. 0 crucibles with covers; Ceramfab pad.

DISCUSSION

The **percentage composition** of a compound is the percentage by mass of each element in the compound. If the formula of a compound is known, the percentage composition can be calculated from the molar mass and the total mass of each element in the compound. The **molar mass** of a compound is determined by adding up the atomic masses of all the atoms making up the formula. The **total mass** of an element in a compound is determined by multiplying the atomic mass of that element by the number of atoms of that element in the formula. The percentage of each element is then calculated by dividing its total mass in the compound by the molar mass of the compound and multiplying by 100.

The percentage composition of many compounds may be directly determined or verified by experimental methods. In this experiment the percentage composition of potassium chlorate will be determined both experimentally and from the formula.

When potassium chlorate is heated to high temperatures (above 400°C) it decomposes to potassium chloride and elemental oxygen, according to the following equation:

$$2\,KClO_3(s) \xrightarrow{\Delta} 2\,KCl(s) + 3\,O_2(g)$$

The relative amounts of oxygen and potassium chloride are measured by heating a weighed sample of potassium chlorate until all of the oxygen has been released from the sample. This is accomplished when the sample is heated to constant weight. In this experiment you will heat, cool, and weigh the sample at least twice. If the sample loses more than 0.05 g after the second heating it has not been heated to constant weight and should be heated a third time.

From the experiment we obtain the following three values:

1. Mass of original sample ($KClO_3$).

2. Mass lost when sample was heated (Oxygen).

3. Mass of residue (KCl).

From these experimental values (and a table of atomic masses) we can calculate the following:

4. Percentage oxygen in sample (Experimental value)

$$= \left(\frac{\text{Mass lost by sample}}{\text{Original sample mass}} \right)(100)$$

5. Percentage KCl in sample (Experimental value)

$$= \left(\frac{\text{Mass of residue}}{\text{Original sample mass}} \right)(100)$$

6. Percentage oxygen in $KClO_3$ from formula (Theoretical value)

$$= \left(\frac{\text{3 at. masses of oxygen}}{\text{Molar mass of } KClO_3} \right)(100) = \left(\frac{3 \times 16.00\,g}{122.6\,g} \right)(100)$$

7. Percentage KCl in $KClO_3$ from formula (Theoretical value)

$$= \left(\frac{\text{Molar mass of KCl}}{\text{Molar mass of } KClO_3} \right)(100) = \left(\frac{74.55\,g}{122.6\,g} \right)(100)$$

8. Percentage error in experimental oxygen determination

$$= \left(\frac{\text{Theoretical value} - \text{Experimental value}}{\text{Theoretical value}} \right)(100)$$

PROCEDURE

 PRECAUTIONS: Since potassium chlorate is a strong oxidizing agent it may cause fires or explosions if mixed or heated with combustible (oxidizable) materials such as paper. Observe the following safety precautions when working with potassium chlorate:

1 **Wear protective glasses.**

2. Use clean crucibles that have been heated and cooled prior to adding potassium chlorate.

3. Use Reagent Grade potassium chlorate.

 4. **Dispose of any excess or spilled potassium chlorate as directed by your instructor. (Potassium chlorate may start fires if mixed with paper or other solid wastes.)**

5. Heat samples slowly and carefully to avoid spattering molten material—and to avoid poor experimental results.

NOTES:

1. Make all weighings to the highest precision possible with the balance available to you. Use the same balance to make all weighings for a given sample. Record all data directly on the report sheet as they are obtained.

2. Duplicate samples of potassium chlorate are to be analyzed, if two crucibles are available.

3. For utmost precision, handle crucibles with tongs after the initial heating.

A. Determining Percentage Composition

Place a clean, dry crucible (uncovered) on a clay triangle and heat for 2 or 3 minutes at the maximum flame temperature. The tip of the sharply defined inner-blue cone of the flame should almost touch and heat the crucible bottom to redness. Allow the crucible to cool. If two crucibles are being used, carefully transfer the first to a Ceramfab pad and heat the second while the first crucible is cooling.

Weigh the cooled crucible and its cover; add between 1 and 1.5 g of potassium chlorate; weigh again.

> **NOTE:** The crucible must be covered when potassium chlorate is being heated in it.

Place the covered crucible on the clay triangle and **heat gently for 8 minutes** with the tip of the inner-blue cone of the flame 6 to 8 cm (about 2.5 to 3 in.) below the crucible bottom. Then carefully lower the crucible or raise the burner until the tip of the sharply defined inner-blue cone just touches the bottom of the crucible, and heat for an additional 10 minutes. The bottom of the crucible should be heated to a dull red color during this period.

Grasp the crucible just below the cover with the concave part of the tongs and very carefully transfer it to a Ceramfab pad. Allow to cool (about 10 minutes) and weigh. Begin analysis of a second sample while the first is cooling.

After weighing, reheat the first sample for an additional 6 minutes at the maximum flame temperature (bottom of the crucible heated to a dull red color); cool and reweigh. If the residue is at constant weight, the last two weighings should be in agreement. If the mass decreased more than 0.05 g between these two weighings, repeat the heating and weighing until two successive weighings agree within 0.05 g. Use the final weight in your calculations.

Complete the analysis of the second sample following the same procedure used for the first.

B. Qualitative Examination of Residue

This part of the experiment should be started as soon as the final heating and weighing of the first sample is completed and while the second sample is in progress.

Number and place three clean test tubes in a rack. Put a pea-sized quantity of potassium chloride into tube No. 1 and a like amount of potassium chlorate into tube No. 2. Add 10 mL of distilled water to each of these two tubes and shake to dissolve the salts. Now add distilled water to the crucible containing the residue from the first sample so it is one-half full. Heat the uncovered crucible very gently for about 1 minute; transfer 1 to 2 mL of the resulting solution from the crucible to tube No. 3; add about 10 mL of distilled water and mix.

Test the solution in each tube as follows: Add 5 drops of dilute (6 M) nitric acid and 5 drops of 0.1 M silver nitrate solution. Mix thoroughly. Record your observations. This procedure using nitric acid and silver nitrate is a general test for chloride ions. The formation of a white precipitation is a positive test and indicates the presence of chloride ions. A positive test is obtained with any substance that produces chloride ions in solution.

 Dispose of solutions and precipitates containing silver in the heavy metal waste container provided. Dispose of the remaining contents in the crucible down the sink.

EXPERIMENT 11

Double Displacement Reactions

MATERIALS AND EQUIPMENT

Solid: sodium sulfite (Na_2SO_3). **Solutions:** dilute (6 M) ammonium hydroxide (NH_4OH), 0.1 M ammonium chloride (NH_4Cl), 0.1 M barium chloride ($BaCl_2$), 0.1 M calcium chloride ($CaCl_2$), 0.1 M copper(II) sulfate ($CuSO_4$), dilute (6 M) hydrochloric acid (HCl), concentrated (12 M) hydrochloric acid (HCl), 0.1 M iron(III) chloride ($FeCl_3$), dilute (6 M) nitric acid (HNO_3), 0.1 M potassium nitrate (KNO_3), 0.1 M silver nitrate ($AgNO_3$), 0.1 M sodium carbonate (Na_2CO_3), 0.1 M sodium chloride (NaCl), 10 percent sodium hydroxide (NaOH), dilute (3 M) sulfuric acid (H_2SO_4), and 0.1 M zinc nitrate [$Zn(NO_3)_2$]. Medicine dropper.

DISCUSSION

Double displacement reactions are among the most common of the simple chemical reactions and are comparatively easy to study.

In each part of this experiment two aqueous solutions, each containing positive and negative ions, will be mixed in a test tube. Consider the hypothetical reaction.

$$AB + CD \longrightarrow AD + CB$$

where $AB(aq)$ exists as A^+ and B^- ions in solution and $CD(aq)$ exists as C^+ and D^- ions in solution. As the ions come in contact with each other, there are six possible combinations that might conceivably cause chemical reaction. Two of these combinations are the meeting of ions of like charge; that is, $A^+ + C^+$ and $B^- + D^-$. But since like charges repel, no reaction will occur. Two other possible combinations are those of the original two compounds; that is, $A^+ + B^-$ and $C^+ + D^-$. Since we originally had a solution containing each of these pairs of ions, they can mutually exist in the same solution; therefore they do not recombine. Thus the two possibilities for chemical reaction are the combination of each of the positive ions with the negative ion of the other compound; that is, $A^+ + D^-$ and $C^+ + B^-$. Let us look at some examples.

Example 1. When solutions of sodium chloride and potassium nitrate are mixed, the equation for the double displacement reaction (hypothetical) is

$$NaCl(aq) + KNO_3(aq) \longrightarrow KCl(aq) + NaNO_3(aq)$$

We get the hypothetical products by simply combining each positive ion with the other negative ion. But has there been a reaction? When we do the experiment, we see no evidence of reaction. There is no precipitate formed, no gas evolved, and no obvious temperature change. Thus we must conclude that no reaction occurred. Both hypothetical products are soluble salts, so the ions are still present in solution. We can say that we simply have a solution of four kinds of ions, Na^+, Cl^-, K^+, and NO_3^-.

The situation is best expressed by changing the equation to

$$NaCl(aq) + KNO_3(aq) \longrightarrow No\ reaction$$

Example 2. When solutions of sodium chloride and silver nitrate are mixed, the equation for the double displacement reaction (hypothetical) is

$$NaCl + AgNO_3 \longrightarrow NaNO_3 + AgCl$$

A white precipitate is produced when these solutions are mixed. This precipitate is definite evidence of a chemical reaction. One of the two products, sodium nitrate ($NaNO_3$) or silver chloride ($AgCl$), is insoluble. Although the precipitate can be identified by further chemical testing, we can instead look at the **Solubility Table in Appendix 5** to find that sodium nitrate is soluble but silver chloride is insoluble. We may then conclude that the precipitate is silver chloride and indicate this in the equation with an (s). Thus

$$NaCl(aq) + AgNO_3(aq) \longrightarrow NaNO_3(aq) + AgCl(s)$$

Example 3. When solutions of sodium carbonate and hydrochloric acid are mixed, the equation for the double displacement reaction (hypothetical) is

$$Na_2CO_3(aq) + 2\,HCl(aq) \longrightarrow 2\,NaCl(aq) + H_2CO_3(aq)$$

Bubbles of a colorless gas are evolved when these solutions are mixed. Although this gas is evidence of a chemical reaction, neither of the indicated products is a gas. But carbonic acid, H_2CO_3, is an unstable compound and readily decomposes into carbon dioxide and water.

$$H_2CO_3(aq) \longrightarrow H_2O(l) + CO_2(g)$$

Therefore, CO_2 and H_2O are the products that should be written in the equation. The original equation then becomes

$$Na_2CO_3(aq) + 2\,HCl(aq) \longrightarrow 2\,NaCl(aq) + H_2O(l) + CO_2(g)$$

The evolution of a gas is indicated by a (g).

Examples of some other substances that decompose to form gases are sulfurous acid (H_2SO_3) and ammonium hydroxide (NH_4OH):

$$H_2SO_3(aq) \longrightarrow H_2O(l) + SO_2(g)$$
$$NH_4OH(aq) \longrightarrow H_2O(l) + NH_3(g)$$

Example 4. When solutions of sodium hydroxide and hydrochloric acid are mixed, the equation for the double displacement reaction (hypothetical) is

$$NaOH(aq) + HCl(aq) \longrightarrow NaCl(aq) + H_2O(l)$$

The mixture of these solutions produces no visible evidence of reaction, but on touching the test tube we notice that it feels warm. The evolution of heat is evidence of a chemical reaction. **Example 4** and **Example 1** appear similar because there is no visible evidence of reaction. However, the difference is very important. In **Example 1** all four ions are still uncombined. In **Example 4** the hydrogen ions (H^+) and hydroxide ions (OH^-) are no longer free in solution but have combined to form water. The reaction of H^+ (an acid) and OH^- (a base) is called **neutralization**. The formation of the slightly ionized compound (water) caused the reaction to occur and was the source of the heat liberated.

Water is the most common slightly ionized substance formed in double displacement reactions; other examples are acetic acid ($HC_2H_3O_2$), oxalic acid ($H_2C_2O_4$), and phosphoric acid (H_3PO_4).

From the four examples cited we see that a double displacement reaction will occur if at least one of the following classes of substances is formed by the reaction:

1. A precipitate

2. A gas

3. A slightly ionized compound, usually water

PROCEDURE

Wear protective glasses.

Each part of the experiment (except No. 12) consists of mixing equal volumes of two solutions in a test tube. Use about a **3 mL sample** of each solution (about 1.5 cm of liquid in a standard test tube). It is not necessary to measure each volume accurately. Record your observations at the time of mixing. Where there is no visible evidence of reaction, feel each tube, or check with a thermometer, to determine if heat is evolved (exothermic reaction). In each case where a reaction has occurred, complete and balance the equation, properly indicating precipitates and gases. When there is no evidence of reaction, write the words "No reaction" as the right-hand side of the equation.

1. Mix 0.1 M sodium chloride and 0.1 M potassium nitrate solutions.

2. Mix 0.1 M sodium chloride and 0.1 M silver nitrate solutions.

3. Mix 0.1 M sodium carbonate and **dilute.** (6 M) hydrochloric acid solutions.

4. Mix 10 percent sodium hydroxide and dil. (6 M) hydrochloric acid solutions.

5. Mix 0.1 M barium chloride and dil. (3 M) sulfuric acid solutions.

 6. Mix **dilute** (6 M) ammonium hydroxide and **dilute** (3 M) sulfuric acid solutions.

7. Mix 0.1 M copper(II) sulfate and 0.1 M zinc nitrate solutions.

8. Mix 0.1 M sodium carbonate and 0.1 M calcium chloride solutions.

9. Mix 0.1 M copper(II) sulfate and 0.1 M ammonium chloride solutions.

10. Mix 10 percent sodium hydroxide and dil. (6 M) nitric acid solutions.

11. Mix 0.1 M iron(III) chloride and dil. (6 M) ammonium hydroxide solutions.

 12. **Do this part in the fume hood.** Add 1 g of solid sodium sulfite to 3 mL of water and shake to dissolve. Now add about 1 mL of conc. (12 M) hydrochloric acid solution, a drop at a time, using a medicine dropper. Observe the results carefully.

 Dispose of mixtures from reactions 2, 5, 7, 9 in the "heavy metal waste" container. Dispose of the contents of reaction 12 in the sink inside the hood. Dispose of the contents of all other tubes in the sink and flush with water.

EXPERIMENT 12

Single Displacement Reactions

MATERIALS AND EQUIPMENT

Solids: strips of sheet copper, lead, and zinc measuring about 1×2 cm; and sandpaper or emery cloth. **Solutions:** 0.1 M copper(II) nitrate [$Cu(NO_3)_2$], 0.1 M lead(II) nitrate [$Pb(NO_3)_2$], 0.1 M magnesium sulfate ($MgSO_4$), 0.1 M silver nitrate ($AgNO_3$), and dilute (3 M) sulfuric acid (H_2SO_4). Small test tubes.

DISCUSSION

The chemical reactivity of elements varies over an immense range. Some, like sodium and fluorine, are so reactive that they are never found in the free or uncombined state in nature. Others, like xenon and platinum, are nearly inert and can be made to react with other elements only under special conditions.

The **reactivity** of an element is related to its tendency to lose or gain electrons; that is, to be oxidized or reduced. In principle it is possible to arrange nearly all the elements into a single series in order of their reactivities. A series of this kind indicates which free elements are capable of displacing other elements from their compounds. Such a list is known as an **activity** or **electromotive series.** To illustrate the preparation of an activity series, we will experiment with a small group of selected elements and their compounds.

A generalized single displacement reaction is represented by the equation

$$A(s) + BC(aq) \longrightarrow B(s) + AC(aq)$$

Element A is the more active element and replaces element B from the compound BC. But if element B is more active than element A, no reaction will occur.

Let us consider two specific examples, using copper and mercury.

Example 1. A few drops of mercury metal are added to a solution of copper(II) chloride ($CuCl_2$).

Example 2. A strip of metallic copper is immersed in a solution of mercury(II) chloride ($HgCl_2$).

In Example 1 no change is observed even after the solution has been standing for a prolonged time, and we conclude that there is no reaction. In Example 2 the copper strip is soon coated with metallic mercury, and the solution becomes pale green. From this evidence we conclude that mercury will not displace copper in copper compounds but copper will displace mercury in mercury compounds. Therefore copper is a more reactive metal than mercury and is above mercury in the activity series. In terms of chemical equations these facts may be represented as

Example 1. $Hg(l) + CuCl_2(aq) \longrightarrow$ No reaction

Example 2. $Cu(s) + HgCl_2(aq) \longrightarrow Hg(l) + CuCl_2(aq)$

The second equation shows that, in terms of oxidation numbers (or charges), the chloride ion remained unchanged, mercury changed from +2 to 0, and copper changed from 0 to +2. The +2 oxidation state of copper is the one normally formed in solution.

Expressed another way, the actual reaction that occurred was the displacement of a mercury ion by a copper atom. This can be expressed more simply in equation form:

$$Cu^0(s) + Hg^{2+}(aq) \longrightarrow Hg^0 + Cu^{2+}(aq)$$

In contrast to double displacement reactions, single displacement reactions involve changes in oxidation numbers and therefore are also classified as **oxidation-reduction reactions.**

PROCEDURE

Wear protective glasses.

1. Place six clean small test tubes in a rack and number them 1–6. To each, add about 2 mL of the solutions listed below.

2. Obtain three pieces of sheet zinc, two of copper, and one of lead. Be sure metal strips are small enough to fit into the test tubes. Clean the metal pieces with fine sandpaper or emery cloth to expose fresh metal surfaces. Add the metals to the test tubes with the solutions as listed.

Tube 1: silver nitrate + copper strip
Tube 2: copper(II) nitrate + lead strip
Tube 3: lead(II) nitrate + zinc strip
Tube 4: magnesium sulfate + zinc strip
Tube 5: dilute (3M) sulfuric acid + copper strip
Tube 6: dilute (3M) sulfuric acid + zinc strip

3. Observe the contents of each tube carefully and record any evidence of chemical reaction.

Evidence of reaction will be either evolution of a gas (bubbles) or appearance of a metallic deposit on the surface of the metal strip. Metals deposited from a solution are often black or gray (in the case of copper, very dark reddish brown) and bear little resemblance to commercially prepared metals.

With some of the combinations used in these experiments, the reactions may be slow or difficult to detect. If you see no immediate evidence of reaction, set the tube aside and allow it to stand for about 10 minutes, then reexamine it.

---

4. Pour the solutions in each test tube into the "heavy metals waste" container. Rinse the metals in tap water and dispose of the strips in the trash. Do not allow the metal strips to go into the sink or into the waste bottle.

EXPERIMENT 14

Identification of Selected Anions

MATERIALS AND EQUIPMENT

Liquids: Decane ($C_{10}H_{22}$). **Solutions:** 0.1 M barium chloride ($BaCl_2$), freshly prepared chlorine water (Cl_2), dilute (6 M) hydrochloric acid (HCl), dilute (6 M) nitric acid (HNO_3), 0.1 M silver nitrate ($AgNO_3$), 0.1 M sodium bromide (NaBr), 0.1 M sodium carbonate (Na_2CO_3), 0.1 M sodium chloride (NaCl), 0.1 M sodium iodide (NaI), 0.1 M sodium phosphate (Na_3PO_4), 0.1 M sodium sulfate (Na_2SO_4), and unknown solutions. Wash bottle for distilled water.

DISCUSSION

The examination of a sample of inorganic material to identify the ions that are present is called **qualitative analysis.** To introduce qualitative analysis, we will analyze for six anions (negatively charged ions). The ions selected for identification are chloride (Cl^-), bromide (Br^-), iodide (I^-), sulfate (SO_4^{2-}), phosphate (PO_4^{3-}) and carbonate (CO_3^{2-}).

Qualitative analysis is based on the fact that no two ions behave identically in all of their chemical reactions. Identification depends on appropriate chemical tests coupled with careful observation of such characteristics as solution color, formation and color of precipitates, evolution of gases, etc. Test reactions are selected to identify the ions in the fewest steps possible. In this experiment only one anion is assumed to be present in each sample. If two or more anions must be detected in a single solution, the scheme of analysis can be considerably more complex.

Silver Nitrate Test

When solutions of the sodium salts of the six anions are reacted with silver nitrate solution, the following precipitates are formed: AgCl, AgBr, AgI, Ag_3PO_4, and Ag_2CO_3. Ag_2SO_4 is moderately soluble and does not precipitate at the concentrations used in these solutions. When dilute nitric acid is added, the precipitates Ag_3PO_4, and Ag_2CO_3 dissolve; AgCl, AgBr, and AgI remain undissolved. Acids react with carbonates to form CO_2 (g). Look for gas bubbles when nitric acid is added to the silver precipitates.

In some cases a tentative identification of an anion may be made from the silver nitrate test. This identification is based on the color of the precipitate and on whether or not the precipitate is soluble in nitric acid. However, since two or more anions may give similar results, second or third confirmatory tests are necessary for positive identification.

Barium Chloride Test

When barium chloride solution is added to solutions of the sodium salts of the six anions, precipitates of $BaSO_4$, $Ba_3(PO_4)_2$ and $BaCO_3$, are obtained. No precipitate is obtained with Cl^-, Br^-, or I^-.

When dilute hydrochloric acid is added, the precipitates $Ba_3(PO_4)_2$ and $BaCO_3$ dissolve; $BaSO_4$ does not dissolve. Look for CO_2 gas bubbles.

Organic Solvent Test

The silver nitrate test can prove the presence of a halide ion (Cl⁻, Br⁻, or I⁻) because the silver precipitates of the other three anions dissolve in nitric acid. But the colors of the three silver halides do not differ sufficiently to establish which halide ion is present.

Adding chlorine water (Cl_2 dissolved in water) to halide salts in solution will oxidize bromide ion to free bromine (Br_2) and iodine ion to free iodine (I_2). The free halogen may be extracted from the water solution by adding an immiscible organic solvent such as decane and shaking vigorously. The colors of the three halogens in organic solvents are quite different. Cl_2 is pale yellow, Br_2 is yellow-orange to reddish-brown, and I_2 is pink to violet. After adding chlorine water and shaking, a yellow-orange to reddish-brown color in the decane layer indicates that Br⁻ was present in the original solution; a pink to violet color in the decane layer indicates that I⁻ was present. However, a pale yellow color does not indicate Cl⁻, since Cl_2 was added as a reagent. But if the silver nitrate test gives a white precipitate that is insoluble in nitric acid, and the organic solvent test shows no Br⁻ or I⁻, then you can conclude that Cl⁻ was present.

Though we have described many of the expected results of these tests, it is necessary to test known solutions to actually see the results of the tests and to develop satisfactory experimental techniques. During this experiment, you will perform these tests on six known anions.

Then, two "unknown" solutions, each containing one of the six anions, will be analyzed. When an unknown is analyzed, the results should agree in all respects with one of the known anions. If the results do not fully agree with one of the six known ions, either the testing has been poorly done or the unknown does not contain any of the specified ions.

Three different kinds of equations may be used to express the behavior of ions in solution. For example, the reaction of the chloride ion (from sodium chloride) may be written.

1. $NaCl(aq) + AgNO_3(aq) \longrightarrow AgCl(s) + NaNO_3(aq)$

2. $Na^+(aq) + Cl^-(aq) + Ag^+(aq) + NO_3^-(aq) \longrightarrow AgCl(s) + Na^+(aq) + NO_3^-(aq)$

3. $Cl^-(aq) + Ag^+(aq) \longrightarrow AgCl(s)$

Equation (1) is the **formula (un-ionized) equation;** it shows the formulas of the substances in the equation as they are normally written. Equation (2) is the **total ionic equation;** it shows the substances as they occur in solution. Strong electrolytes are written as ions; weak electrolytes, precipitates, and gases are written in their un-ionized or molecular form. Equation (3) is the **net ionic equation;** it includes only those substances or ions in Equation (2) that have undergone a chemical change. Thus Na^+ and NO_3^- (sometimes called the "spectator" ions) have not changed and do not appear in the net ionic equation. In both the total ionic and net ionic equations, the atoms and charges must be balanced.

PROCEDURE

Wear protective glasses

1. Clean eight test tubes and rinse each twice with 5 mL of distilled water. The first six test tubes are for the known solutions that will be tested to demonstrate the expected reactions with each anion. Use a marker to label these tubes as follows: NaCl, NaBr, NaI, Na_2SO_4, Na_3PO_4 and Na_2CO_3. The last two tubes are for your unknowns and should be left blank for now. Arrange these test tubes in order in your test tube rack.

2. Clean and rinse two more test tubes and take them to your instructor for your unknown solutions and their identification code. Label them with the code numbers immediately. To avoid possible confusion with the empty unknown test tubes in the rack, put these coded tubes aside in a beaker. Record the code of these unknowns in the top right-hand columns of your report form and label each of the blank tubes in the rack with one of these unknown code numbers.

Pour 2 mL (no more) of each of the six known solutions—one solution per tube—and 2 mL of the corresponding unknown into each unknown tube. Save the remaining portions of the unknown solutions for tests B and C.

You can save considerable time by measuring out 2 mL into the first test tube and using the height of this liquid in the test tube as a guide for measuring out the others.

 Dispose of solutions containing decane in the container marked "Waste organic solvents." Dispose of solutions containing silver, and barium, in the "heavy metals waste" container.

For each of the following tests that will be performed on known and unknown solutions, there is a corresponding block on the report form where observations should be recorded. If a precipitate forms, record "ppt formed" and include its color. If no precipitate forms, record "no ppt." When dissolving precipitates, record "ppt dissolved" or "ppt did not dissolve." For the decane solubility test, indicate the color of the decane layer.

A. Silver Nitrate Test

 Silver nitrate will stain your skin black. If any silver nitrate gets on your hands, wash it off immediately to avoid these stains.

Add about 1 mL of 0.1 M silver nitrate solution to each test tube. Record the results. Now add about 3 mL of dilute (6 M) nitric acid to each test tube; stopper and shake well. Record the results.

B. Barium Chloride Test

Wash all eight test tubes and rinse each tube twice with distilled water. Again put about 2 mL of the specified solution into each of the eight test tubes. Add about 2 mL of 0.1 M barium chloride solution to each test tube and mix. Record the results. Now add 3 mL of dilute hydrochloric acid to each tube; stopper and shake well. Record the results.

C. Organic Solvent Test

Again wash and rinse all eight test tubes. Again put about 2 mL of the specified solution into each of the eight test tubes. Now add about 2 mL of decane and about 2 mL of chlorine water to each test tube; stopper and shake well. Record the results.

After completing the three tests, compare the results of the known solutions with your observations for your unknown solutions. Record the formula of the anion present in each solution on the report form (Part D).

REPORT FOR EXPERIMENT 14

Identification of Selected Anions

	NaCl	NaBr	NaI	Na_2SO_4	Na_3PO_4	Na_2CO_3	Unknown No. ____	Unknown No. ____
A. AgNO₃ Test Addition of AgNO₃ solution								
Addition of dil. HNO₃								
B. BaCl₂ Test Addition of BaCl₂ solution								
Addition of dil. HCL								
C. Organic Solvent Test Color of decane layer								
D. Formula of anion present in the solution tested.								

QUESTIONS AND PROBLEMS

1. The following three solutions were analyzed according to the scheme used in this experiment. Which one, if any, of the ions tested, is present in each solution? If the data indicate that none of the six is present, write the word "None" as your answer.

(a) **Silver Nitrate Test.**   Yellow precipitate formed, which dissolved in dilute nitric acid.

   Barium Chloride Test.   White precipitate formed, which dissolved in dilute hydrochloric acid.

   Organic Solvent Test.   The decane layer remained almost colorless after treatment with chlorine water.

Anion present _____

(b) **Silver Nitrate Test.**   Red precipitate formed, which dissolved in dilute nitric acid to give an orange solution.

   Barium Chloride Test.   Yellow precipitate formed, which dissolved in dilute hydrochloric acid to give an orange solution.

   Organic Solvent Test.   The decane layer remained almost colorless after treatment with chlorine water.

Anion present _____

(c) **Silver Nitrate Test.**   Yellow precipitate formed, which did not dissolve in dilute nitric acid.

   Barium Chloride Test.   No precipitate formed.

   Organic Solvent Test.   The decane layer turned reddish-brown.

Anion present _____

2. Write formula, total ionic, and net ionic equations for the following reactions: Use the solubility table in Appendix 5 for reactions that were not observed directly in this experiment. All reactions are in aqueous solutions.

(a) Sodium bromide and silver nitrate.

(b) Sodium carbonate and silver nitrate.

(c) Sodium arsenate and barium chloride.

3. Write net ionic equations for the following reactions. Assume that a precipitate is formed in each case.

(a) Sodium iodide and silver nitrate.

(b) Sodium acetate and silver nitrate.

(c) Sodium phosphate and barium chloride.

(d) Sodium sulfate and barium chloride.

EXPERIMENT 22

Neutralization–Titration I

MATERIALS AND EQUIPMENT

Solid: potassium hydrogen phthalate, abbreviated KHP ($KHC_8H_4O_4$). **Liquids:** phenolphthalein indicator, unknown base solution (NaOH). One buret (25 mL or 50 mL) and buret clamp, buret brush. Wash bottle for distilled water.

DISCUSSION

The reaction of an acid and a base to form a salt and water is known as **neutralization.** In this experiment potassium hydrogen phthalate (abbreviated KHP) is used as the acid. Potassium hydrogen phthalate is an organic substance having the formula $HKC_8H_4O_4$, and like HCl, has only one acid hydrogen atom per molecule. Because of its complex formula, potassium hydrogen phthalate is commonly called KHP Despite its complex formula we see that the reaction of KHP with sodium hydroxide is similar to that of HCl. One mole of KHP reacts with one mole of NaOH.

$$HKC_8H_4O_4 + NaOH \longrightarrow NaKC_8H_4O_4 + H_2O$$

$$HCl + NaOH \longrightarrow NaCl + H_2O$$

Titration is the process of measuring the volume of one reagent required to react with a measured volume or mass of another reagent. In this experiment we will determine the molarity of a base (NaOH) solution from data obtained by titrating KHP with the base solution. The base solution is added from a buret to a flask containing a weighed sample of KHP dissolved in water. From the mass of KHP used we calculate the moles of KHP. Exactly the same number of moles of base is needed to neutralize this number of moles of KHP since one mole of NaOH reacts with one mole of KHP. We then calculate the molarity of the base solution from the titration volume and the number of moles of NaOH in that volume.

In the titration, the point of neutralization, called the **end-point,** is observed when an indicator, placed in the solution being titrated, changes color. The indicator selected is one that changes color when the stoichiometric quantity of base (according to the chemical equation) has been added to the acid. A solution of phenolphthalein, an organic acid, is used as the indicator in this experiment. Phenolphthalein is colorless in acid solution but changes to pink when the solution becomes slightly alkaline. When the number of moles of sodium hydroxide added is equal to the number of moles of KHP originally present, the reaction is complete. The next drop of sodium hydroxide added changes the indicator from colorless to pink.

Use the following relationships in your calculations:

1. According to the equation for the reaction,

   Moles of KHP reacted = Moles of NaOH reacted

2. $\text{Moles} = \dfrac{\text{g of solute}}{\text{molar mass of solute}}$

3. Molarity is an expression of concentration, the units of which are moles of solute per liter of solution:

$$\text{Molarity} = \frac{\text{moles}}{\text{liter}}$$

Thus, a 1.00 molar (1.00 M) solution contains 1.00 mole of solute in 1 liter of solution. A 0.100 M solution, then, contains 0.100 mole of solute in 1 liter of solution.

4. The number of moles of solute present in a known volume of solution of known concentration can be calculated by multiplying the volume of the solution (in liters) by the molarity of the solution:

$$\text{Moles} = (\text{liters})(\text{molarity}) = (\text{liters})\left(\frac{\text{moles}}{\text{liter}}\right)$$

PROCEDURE

Wear protective glasses.

 Dispose of all solutions in the sink.

Make all weighings to the highest precision of the balance.

Obtain some solid KHP in a test tube or vial. Weigh two samples of KHP into 125 mL Erlenmeyer flasks, numbered for identification. (The flasks should be rinsed with distilled water, but need not be dry on the inside.) First weigh the flask, then add KHP to the flask by tapping the test tube or vial until 1.000 to 1.200 g has been added (see Figure 22.1). Determine the mass of the flask and the KHP. In a similar manner weigh another sample of KHP into the second flask. To each flask add approximately 30 mL of distilled water. If some KHP is sticking to the walls of the flask, rinse it down with water from a wash bottle. Warm the flasks slightly and swirl them until all the KHP is dissolved.

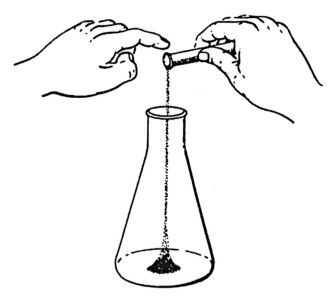

Figure 22.1   Method of adding KHP from a vial to a weighed Erlenmeyer flask

Obtain one buret and clean it. See "Use of the Buret," on the following page for instructions on cleaning and using the buret. Read and record all buret volumes to the nearest 0.01 mL.

Obtain about 250 mL of a base (NaOH) of unknown molarity in a clean, **dry** 250 mL Erlenmeyer flask as directed by your instructor. Record the number of this unknown.

1. Keep your base solution stoppered when not in use.

2. The 250 mL sample of base is intended to be used in both this experiment and Experiment 23. Be sure to label and save it.

Rinse the buret with two 5 to 10 mL portions of the base, running the second rinsing through the buret tip. Discard the rinsings in the sink. Fill the buret with the base, making sure that the tip is completely filled and contains no air bubbles. Adjust the level of the liquid in the buret so that the bottom of the meniscus is at exactly 0.00 mL. Record the initial buret reading (0.00 mL) in the space provided on the report form.

Add 3 drops of phenolphthalein solution to each 125 mL flask containing KHP and water. Place the first (Sample 1) on a piece of white paper under the buret extending the tip of the buret into the flask (see Figure 22.2).

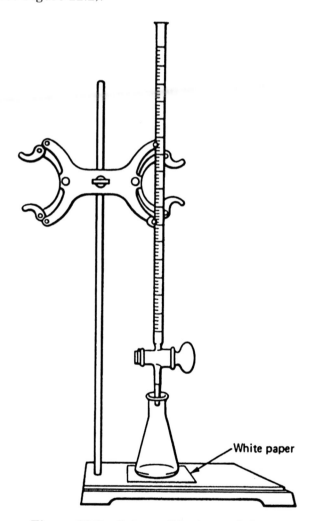

Figure 22.2   Setup with stopcock buret

Titrate the KHP by adding base until the end-point is reached. The titration is conducted by swirling the solution in the flask with the right hand (if you are right handed) while manipulating the stopcock with the left (Figure 22.3). As base is added you will observe a pink color caused by localized high base concentration. Toward the end-point the color flashes throughout the solution, remaining for a longer time. When this occurs, add the base drop by drop until the end-point is reached, as indicated by the first drop of base which causes a faint pink color to remain in the entire solution for at least 30 seconds. Read and record the final buret reading (see Figure 22.5). Refill the buret to the zero mark and repeat the titration with Sample 2. Then, calculate the molarity of the base in each sample. If these molarities differ by more than 0.004, titrate a third sample.

When you are finished with the titrations, empty and rinse the buret at least twice (including the tip) with tap water and once with distilled water. Return the vial with the unused KHP.

Use of the Buret

A buret is a volumetric instrument that is calibrated to deliver a measured volume of solution. The 50 mL buret is calibrated from 0 to 50 mL in 0.1 mL increments and is read to the nearest 0.01 mL. All volumes delivered from the buret should be between the calibration marks. (Do not estimate above the 0 mL mark or below the 50 mL mark.)

1. **Cleaning the Buret.** The buret must be clean in order to deliver the calibrated volume. Drops of liquid clinging to the sides as the buret is drained are evidence of a dirty buret.

To clean the buret, first rinse it a couple of times with tap water, pouring the water from a beaker. Then scrub it with a detergent solution, using a long-handled buret brush. Rinse the buret several times with tap water and finally with distilled water. Check for cleanliness by draining the distilled water through the tip and observe whether droplets of water remain on the inner walls of the buret.

2. **Using the Buret.** After draining the distilled water, rinse the buret with two 5 to 10 mL portions of the titrating solution to be used in it. This rinsing is done by holding the buret in a horizontal position and rolling the solution around to wet the entire inner surface. Allow the final rinsing to drain through the tip.

Fill the buret with the solution to slightly above the 0 mL mark and adjust it to 0.00 mL, or some other volume below this mark, by draining the solution through the tip. The buret tip must be completely filled to deliver the volume measured.

To deliver the solution from the buret, turn the stopcock with the forefinger and the thumb of your left hand (if you are right handed) to allow the solution to enter the flask. (See Figure 22.3). This procedure leaves your right hand free to swirl the solution in the flask during the titration. With a little practice you can control the flow so that increments as small as 1 drop of solution can be delivered.

3. **Reading the Buret.** The smallest calibration mark of a 50 mL buret is 0.1 mL. However, the buret is read to the nearest 0.01 mL by estimating between the calibration marks. When reading the buret be sure your line of sight is level with the bottom of the meniscus in order to avoid parallax errors (see Figure 22.4). The exact bottom of the meniscus may be made more prominent and easier to read by allowing the meniscus to pick up the reflection from a heavy dark line on a piece of paper (see Figure 22.5).

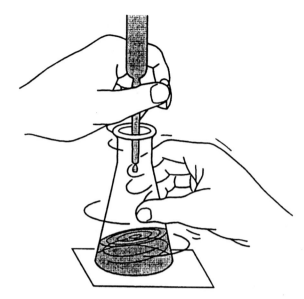

Figure 22.3 Titration technique

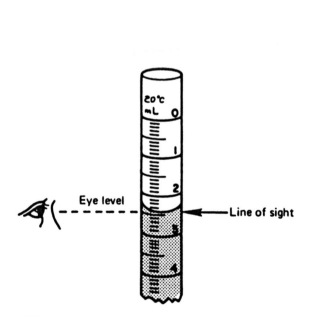

Figure 22.4 Reading the buret. The line of sight must be level with the bottom of the meniscus to avoid parallax.

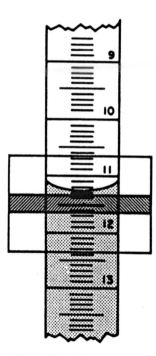

Figure 22.5 Reading the meniscus. A heavy dark line brought to within one division of the meniscus will make the meniscus more prominent and easier to read. The volume reading is 11.28 mL.

EXPERIMENT 23

Neutralization – Titration II

MATERIALS AND EQUIPMENT

Solutions: Acid of unknown molarity, standard base solution (NaOH), vinegar, phenolphthalein indicator. Suction bulb, buret, buret brush, buret clamp, 10 mL volumetric pipet. Wash bottle for distilled water.

DISCUSSION

This experiment may follow Experiment 22 or it may be completed independently of Experiment 22. In either case the discussion section of Experiment 22 supplements the following discussion.

The reaction of an acid and a base to form water and a salt is known as **neutralization.** Hydrochloric acid and sodium hydroxide, for example, react to form sodium chloride and water.

$$HCl(aq) + NaOH(aq) \longrightarrow H_2O(l) + NaCl(aq)$$

The ionic reaction in neutralizations of this type is that of hydrogen (or hydronium) ion reacting with hydroxide ion to form water.

$$H^+(aq) + OH^-(aq) \longrightarrow H_2O(l) \quad \text{or} \quad H_3O^+(aq) + OH^-(aq) \longrightarrow 2\,H_2O(l)$$

A monoprotic acid—i.e., an acid having one ionizable hydrogen atom per molecule—reacts with sodium hydroxide (or any other monohydroxy base) on a 1:1 mole basis. This fact is often utilized in determining the concentrations of solutions of acids by titration.

Titration is the process of measuring the volume of one reagent to react with a measured volume or mass of another reagent. In this experiment an acid solution of unknown concentration is titrated with a base solution of known concentration, Phenolphthalein is used as an indicator. This substance is colorless in acid solution, but changes to pink when the solution becomes slightly basic or alkaline. The change of color, caused by a single drop of the base solution in excess over that required to neutralize the acid, marks the **end-point** of the titration.

Molarity (M) is the concentration of a solution expressed in terms of moles of solute per liter of solution.

$$\text{Molarity} = \frac{\text{moles}}{\text{liter}}$$

Thus a solution containing 1.00 mole of solute in 1.00 liter of solution is 1.00 molar (1.00 M). If only 0.155 mole is present in 1.00 liter of solution, it is 0.155 M, etc. To determine the molarity of any quantity it is only necessary to divide the total number of moles of solute present in the solution by the volume (in liters).

To determine the number of moles of solute present in a known volume of solution, multiply the volume in liters by the molarity.

$$\text{Moles} = (\text{liters})(\text{molarity}) = (\text{liters})\left(\frac{\text{moles}}{\text{liter}}\right)$$

For titrations involving monoprotic acids and monohydroxy bases (one hydroxide ion per formula unit), the number of moles of acid is identical to the number of moles of base required to neutralize the acid. In this experiment we measure the volume of base of known molarity required to neutralize a measured volume of acid of unknown molarity. The molarity of the acid can then be calculated.

$$\text{Moles base} = (\text{liters})(\text{molarity}) = (\text{liters base})\left(\frac{\text{moles base}}{\text{liters}}\right)$$

$$\text{Moles acid} = (\text{moles base})\left(\frac{1 \text{ mole acid}}{1 \text{ mole base}}\right)$$

$$\text{Molarity of acid} = \frac{\text{moles acid}}{\text{liters acid}}$$

In order to determine the molarity of an acid solution, it is not actually necessary to know what the acid is—only whether it is monoprotic, diprotic, or triprotic. The calculations in this experiment are based on the assumption that the acid in the unknown is monoprotic.

If the molarity and the formula of the solute are known, the concentration in grams of solute per liter of the solution may be calculated by multiplying by the molar mass.

$$(\text{Molarity})(\text{molar mass}) = \left(\frac{\text{moles}}{\text{liter}}\right)\left(\frac{\text{grams}}{\text{mole}}\right) = \frac{\text{grams}}{\text{liter}}$$

In determining the acid content of commercial vinegar, it is customary to treat the vinegar as a dilute solution of acetic acid, $HC_2H_3O_2$. The acetic acid concentration of the vinegar may be calculated as grams of acetic acid per liter or as percent acid by mass. If the acetic acid content is to be expressed on a mass percent basis, the density of the vinegar must also be known.

PROCEDURE

Wear protective glasses.

 Do not pipet by mouth.

 Dispose of all solutions in the sink. Flush with water.

A. Molarity of an Unknown Acid

Obtain a sample of acid of unknown molarity in a clean, dry 125 mL Erlenmeyer flask as directed by your instructor.

With a volumetric pipet, transfer a 10.00 mL sample of the acid to a clean, but not necessarily dry, Erlenmeyer flask. See "Use of the Pipet," on the following page, for instructions on cleaning and using the pipet. Pipet a duplicate 10.00 mL sample into a second flask. (If pipets are not available, a buret which has been carefully cleaned and rinsed may be used to measure the acid samples.)

You will need about 150 mL of base of known molarity (standard solution). Your instructor will give you the exact molarity of the base solution that you used in Experiment 22 or you may be given another sodium hydroxide solution of known molarity. Record the exact molarity of this solution. Keep the flask containing the base stoppered when not in use.

Clean and set up a buret. See "Use of the Buret," in Experiment 22, for instructions on cleaning and using the buret.

Rinse the buret with two 5 to 10 mL portions of the base, running the second rinsing through the buret tip. Discard the rinsings in the sink. Fill the buret with the base, making sure that the tip is completely filled and contains no air bubbles. Adjust the level of the liquid in the buret so that the bottom of the meniscus is near or exactly at 0.00 mL. Record the initial buret reading in the space provided on the report form.

Add three drops of phenolphthalein solution and about 25 mL of distilled water to the flask containing the 10.00 mL of acid. Place this flask on a piece of white paper under the buret and lower the buret tip into the flask (see Figure 22.2).

Titrate the acid by adding base until the end-point is reached. During the titration swirl the solution in the flask with the right hand (if you are right handed) while manipulating the stopcock with the left. As the base is added you will observe a pink color caused by localized high base concentration. Near the end-point this color flashes throughout the solution, remaining for increasingly longer periods of time. When this occurs, add the base drop by drop until the end-point is reached, as indicated by the first drop of base which causes the entire solution to retain a faint pink color for at least 30 seconds. Record the final buret reading.

Refill the buret with base and adjust the volume to near the zero mark. Titrate the duplicate sample of acid. If the volumes of base used differ by more than 0.20 mL, titrate a third sample. In the calculations, assume that the unknown acid reacts like KHP or HCl (one mole of acid reacts with one mole of base).

B. Acetic Acid Content of Vinegar

Obtain about 40 mL of vinegar in a clean, dry 50 mL beaker. Record the sample number, if any, of this vinegar.

Titrate duplicate 10.00 mL samples of vinegar using exactly the same procedure outlined in Part A. Remember to rinse the pipet with vinegar before pipeting the vinegar samples.

When you are finished with the titrations, empty the buret and rinse it and the pipet at least twice with tap water and once with distilled water.

Use of the Pipet

A volumetric (transfer) pipet (Figure 23.1) is calibrated to deliver a specified volume of liquid to a precision of about ± 0.02 mL in a 10 mL pipet. To achieve this precision, the pipet must be clean and used in a specified manner.

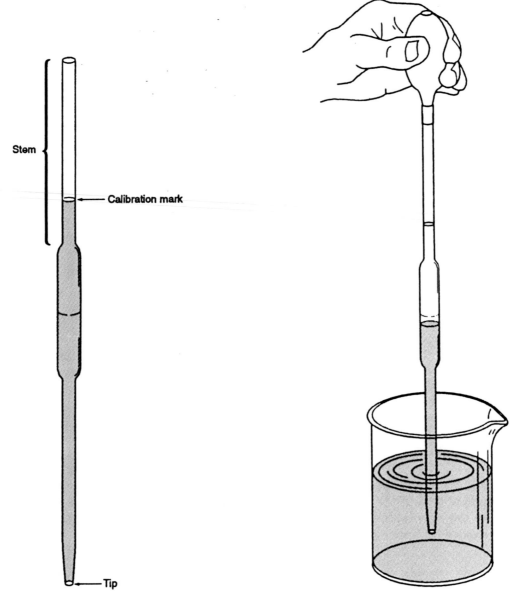

Figure 23.1   A volumetric (transfer) pipet

Stem

Calibration mark

Tip

Figure 23.2   Liquid is drawn into the pipet with a rubber suction bulb. Keep the tip of the pipet below the liquid level during suction.

Liquids are drawn into a pipet by means of a rubber suction bulb (Figure 23.2) or by a rubber tube connected to a water aspirator pump. Suction by mouth has also been used to draw liquids into a pipet, but this is a dangerous practice and is not recommended.

1. **Cleaning the Pipet.**   Use a rubber suction bulb to draw up enough detergent solution to fill about two-thirds of the body or bulb of the pipet. Retain this solution in the pipet by pressing the forefinger tightly against the top of the pipet stem (Figure 23.3). turn the pipet to a nearly horizontal position and gently shake and rotate it until the entire inside surface is wetted. Allow the pipet to drain and rinse it at least three times with tap water and once with distilled water.

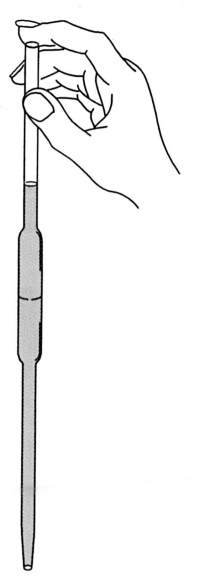

Figure 23.3   Liquid is retained in the pipet by applying pressure with the forefinger to the top of the stem.

Figure 23.4   The pipet is calibrated to deliver the specified volume, leaving a small amount of liquid in the tip.

2. **Using the Pipet.**   Unless the pipet is known to be clean and absolutely dry on the inside, it must be rinsed twice with small portions of the liquid that is to be pipeted. This is done as in the washing procedure described above. These rinses are discarded in order to avoid contamination of the liquid being pipeted. A **pipet** does not need to be rinsed between successive pipettings of the same solution.

To transfer a measured volume of a liquid, collapse a suction bulb by squeezing and place it tightly against the top of a pipet. (Do not try to push the bulb on to the pipet.) Draw the liquid into the pipet until it is filled to about 5 cm above the calibration mark by allowing the bulb to slowly expand. Be careful—do not allow the liquid to get into the bulb. Remove the bulb and quickly place your forefinger over the top of the pipet stem. The liquid will be retained in the pipet if the finger is pressed tightly against the top of the stem. Keeping the pipet in a vertical position, decrease the finger pressure very slightly, and allow the liquid level to drop slowly toward the calibration mark. When the liquid level has almost reached the calibration mark, again increase the finger pressure and stop the liquid when the bottom of the meniscus is exactly on the calibration mark. Touch the tip to the wall of the flask to remove the adhering drop of liquid.

Move the pipet to the flask which is to receive the sample and allow the liquid to drain while holding the pipet in a vertical position. About 10 seconds after the liquid has stopped running from the pipet, touch the tip to the inner wall of the sample flask to remove the drop of liquid adhering to the tip. A small amount of liquid will remain in the tip (Figure 23.4). Do not blow or shake this liquid into the sample; the pipet is calibrated to deliver the volume specified without this small residual.

If you have never used a volumetric pipet, it is advisable to practice by pipetting some samples of distilled water until you have mastered the technique.

APPENDIX 5

Solubility Table

	$C_2H_3O_2^-$	AsO_4^{3-}	Br^-	CO_3^{2-}	Cl^-	CrO_4^{2-}	OH^-	I^-	NO_3^-	$C_2O_4^{2-}$	O^{2-}	PO_4^{3-}	SO_4^{2-}	S^{2-}	SO_3^{2-}
Al^{3+}	aq	I	aq	–	aq	–	I	aq	aq	–	I	I	aq	d	–
NH_4^+	aq	aq	aq	aq	aq	aq	aq	aq	aq	aq	–	aq	aq	aq	aq
Ba^{2+}	aq	I	aq	I	aq	I	sl. aq	aq	aq	I	sl. aq	I	I	d	I
Bi^{3+}	–	sl. aq	d	I	d	–	I	I	d	I	I	sl. aq	d	I	–
Ca^{2+}	aq	I	aq	I	aq	aq	I	aq	aq	I	I	I	I	d	I
Co^{2+}	aq	I	aq	I	aq	I	I	aq	aq	I	I	I	aq	I	I
Cu^{2+}	aq	I	aq	I	aq	I	I	–	aq	I	I	I	aq	I	–
Fe^{2+}	aq	I	aq	sl. aq	aq	–	I	aq	aq	I	I	I	aq	I	sl. aq
Fe^{3+}	I	I	aq	I	aq	I	I	–	aq	aq	I	I	aq	I	–
Pb^{2+}	aq	I	I	I	I	I	I	I	aq	I	I	I	I	I	I
Mg^{2+}	aq	d	aq	I	aq	aq	I	aq	aq	I	I	I	aq	d	sl. aq
Hg_2^{2+}	sl. aq	I	I	I	I	sl. aq	–	I	aq	I	I	I	I	I	–
Hg^{2+}	aq	I	I	I	aq	sl. aq	I	I	aq	I	I	I	d	I	–
K^+	aq	aq	aq	aq	aq	aq	aq	aq	aq	aq	aq	aq	aq	aq	aq
Ag^+	sl. aq	I	I	I	I	I	–	I	aq	I	I	I	I	I	I
Na^+	aq	aq	aq	aq	aq	aq	aq	aq	aq	aq	aq	aq	aq	aq	aq
Zn^{2+}	aq	I	aq	I	aq	I	I	aq	aq	I	I	I	aq	I	I

Key: aq = Soluble in water I = Insoluble in water (less than 1 g/100 g H_2O)
sl. aq = Slightly soluble in water d = Decomposes in water

Periodic Table of the Elements

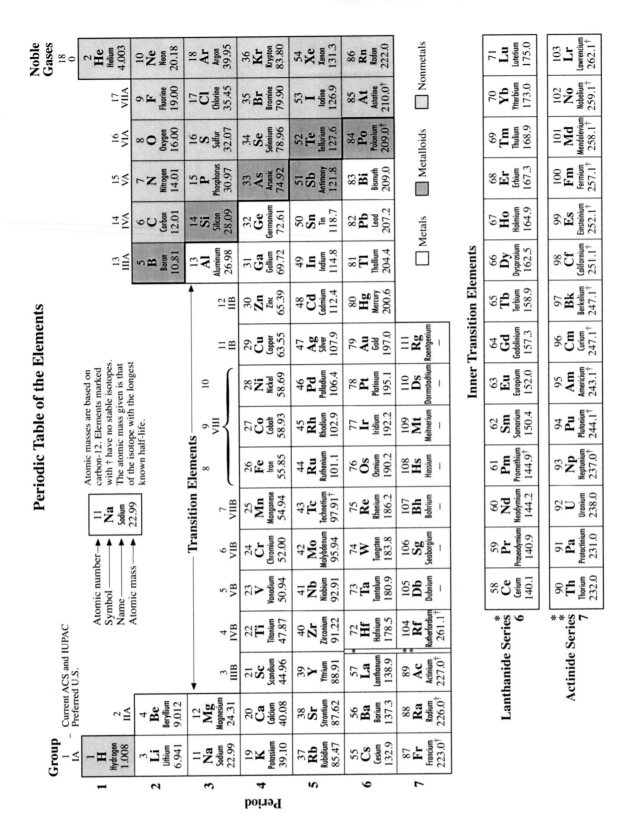

Group

Period	1 IA	2 IIA	3 IIIB	4 IVB	5 VB	6 VIB	7 VIIB	8	9 VIII	10	11 IB	12 IIB	13 IIIA	14 IVA	15 VA	16 VIA	17 VIIA	18 0
1	1 **H** Hydrogen 1.008																	2 **He** Helium 4.003
2	3 **Li** Lithium 6.941	4 **Be** Beryllium 9.012											5 **B** Boron 10.81	6 **C** Carbon 12.01	7 **N** Nitrogen 14.01	8 **O** Oxygen 16.00	9 **F** Fluorine 19.00	10 **Ne** Neon 20.18
3	11 **Na** Sodium 22.99	12 **Mg** Magnesium 24.31											13 **Al** Aluminum 26.98	14 **Si** Silicon 28.09	15 **P** Phosphorus 30.97	16 **S** Sulfur 32.07	17 **Cl** Chlorine 35.45	18 **Ar** Argon 39.95
4	19 **K** Potassium 39.10	20 **Ca** Calcium 40.08	21 **Sc** Scandium 44.96	22 **Ti** Titanium 47.87	23 **V** Vanadium 50.94	24 **Cr** Chromium 52.00	25 **Mn** Manganese 54.94	26 **Fe** Iron 55.85	27 **Co** Cobalt 58.93	28 **Ni** Nickel 58.69	29 **Cu** Copper 63.55	30 **Zn** Zinc 65.39	31 **Ga** Gallium 69.72	32 **Ge** Germanium 72.61	33 **As** Arsenic 74.92	34 **Se** Selenium 78.96	35 **Br** Bromine 79.90	36 **Kr** Krypton 83.80
5	37 **Rb** Rubidium 85.47	38 **Sr** Strontium 87.62	39 **Y** Yttrium 88.91	40 **Zr** Zirconium 91.22	41 **Nb** Niobium 92.91	42 **Mo** Molybdenum 95.94	43 **Tc** Technetium 97.91†	44 **Ru** Ruthenium 101.1	45 **Rh** Rhodium 102.9	46 **Pd** Palladium 106.4	47 **Ag** Silver 107.9	48 **Cd** Cadmium 112.4	49 **In** Indium 114.8	50 **Sn** Tin 118.7	51 **Sb** Antimony 121.8	52 **Te** Tellurium 127.6	53 **I** Iodine 126.9	54 **Xe** Xenon 131.3
6	55 **Cs** Cesium 132.9	56 **Ba** Barium 137.3	57 **La*** Lanthanum 138.9	72 **Hf** Hafnium 178.5	73 **Ta** Tantalum 180.9	74 **W** Tungsten 183.8	75 **Re** Rhenium 186.2	76 **Os** Osmium 190.2	77 **Ir** Iridium 192.2	78 **Pt** Platinum 195.1	79 **Au** Gold 197.0	80 **Hg** Mercury 200.6	81 **Tl** Thallium 204.4	82 **Pb** Lead 207.2	83 **Bi** Bismuth 209.0	84 **Po** Polonium 209.0†	85 **At** Astatine 210.0†	86 **Rn** Radon 222.0
7	87 **Fr** Francium 223.0†	88 **Ra** Radium 226.0†	89 **Ac**** Actinium 227.0†	104 **Rf** Rutherfordium 261.1†	105 **Db** Dubnium –	106 **Sg** Seaborgium –	107 **Bh** Bohrium –	108 **Hs** Hassium –	109 **Mt** Meitnerium –	110 **Ds** Darmstadtium –	111 **Rg** Roentgenium –							

Transition Elements

Atomic masses are based on carbon-12. Elements marked with † have no stable isotopes. The atomic mass given is that of the isotope with the longest known half-life.

Atomic number → 11
Symbol → **Na**
Name → Sodium
Atomic mass → 22.99

– Current ACS and IUPAC
– Preferred U.S.

☐ Metals
▨ Metalloids
▨ Nonmetals

Inner Transition Elements

Lanthanide Series * 6

58 **Ce** Cerium 140.1	59 **Pr** Praseodymium 140.9	60 **Nd** Neodymium 144.2	61 **Pm** Promethium 144.9†	62 **Sm** Samarium 150.4	63 **Eu** Europium 152.0	64 **Gd** Gadolinium 157.3	65 **Tb** Terbium 158.9	66 **Dy** Dysprosium 162.5	67 **Ho** Holmium 164.9	68 **Er** Erbium 167.3	69 **Tm** Thulium 168.9	70 **Yb** Ytterbium 173.0	71 **Lu** Lutetium 175.0

Actinide Series ** 7

90 **Th** Thorium 232.0	91 **Pa** Protactinium 231.0	92 **U** Uranium 238.0	93 **Np** Neptunium 237.0†	94 **Pu** Plutonium 244.1†	95 **Am** Americium 243.1†	96 **Cm** Curium 247.1†	97 **Bk** Berkelium 247.1†	98 **Cf** Californium 251.1†	99 **Es** Einsteinium 252.1†	100 **Fm** Fermium 257.1†	101 **Md** Mendelevium 258.1†	102 **No** Nobelium 259.1†	103 **Lr** Lawrencium 262.1†

Atomic Masses of the Elements
Based on the IUPAC Table of Atomic Masses

Name	Symbol	Atomic Number	Atomic Mass	Name	Symbol	Atomic Number	Atomic Mass
Actinium*	Ac	89	227.0277	Mendelevium*	Md	101	258.0984
Aluminum	Al	13	26.981538	Mercury	Hg	80	200.59
Americium*	Am	95	243.0614	Molybdenum	Mo	42	95.94
Antimony	Sb	51	121.760	Neodymium	Nd	60	144.24
Argon	Ar	18	39.948	Neon	Ne	10	20.1797
Arsenic	As	33	74.92160	Neptunium*	Np	93	237.0482
Astatine*	At	85	209.9871	Nickel	Ni	28	58.6934
Barium	Ba	56	137.327	Niobium	Nb	41	92.90638
Berkelium*	Bk	97	247.0703	Nitrogen	N	7	14.00674
Beryllium	Be	4	9.012182	Nobelium*	No	102	259.1011
Bismuth	Bi	83	208.98038	Osmium	Os	76	190.23
Bohrium*	Bh	107	—	Oxygen	O	8	15.9994
Boron	B	5	10.811	Palladium	Pd	46	106.42
Bromine	Br	35	79.904	Phosphorus	P	15	30.973762
Cadmium	Cd	48	112.411	Platinum	Pt	78	195.078
Calcium	Ca	20	40.078	Plutonium*	Pu	94	244.0642
Californium*	Cf	98	251.0796	Polonium*	Po	84	208.9824
Carbon	C	6	12.0107	Potassium	K	19	39.0983
Cerium	Ce	58	140.116	Praseodymium	Pr	59	140.90765
Cesium	Cs	55	132.90545	Promethium*	Pm	61	144.9127
Chlorine	Cl	17	35.4527	Protactinium	Pa	91	231.03588
Chromium	Cr	24	51.9961	Radium*	Ra	88	226.0245
Cobalt	Co	27	58.933200	Radon*	Rn	86	222.0176
Copper	Cu	29	63.546	Rhenium	Re	75	186.207
Curium*	Cm	96	247.0703	Rhodium	Rh	45	102.90550
Darmstadtium*	Ds	110	—	Roentgenium	Rg	111	—
Dubnium*	Db	105	—	Rubidium	Rb	37	85.4678
Dysprosium	Dy	66	162.50	Ruthenium	Ru	44	101.07
Einsteinium*	Es	99	252.0830	Rutherfordium	Rf	104	261.1089
Erbium	Er	68	167.26	Samarium	Sm	62	150.36
Europium	Eu	63	151.964	Scandium	Sc	21	44.955910
Fermium*	Fm	100	257.0951	Seaborgium	Sg	106	—
Fluorine	F	9	18.9984032	Selenium	Se	34	78.96
Francium*	Fr	87	233.0197	Silicon	Si	14	28.0855
Gadolinium	Gd	64	157.25	Silver	Ag	47	107.8682
Gallium	Ga	31	69.723	Sodium	Na	11	22.989770
Germanium	Ge	32	72.61	Strontium	Sr	38	87.62
Gold	Au	79	196.96655	Sulfur	S	16	32.066
Hafnium	Hf	72	178.49	Tantalum	Ta	73	180.9479
Hassium*	Hs	108	—	Technetium*	Tc	43	97.09072
Helium	He	2	4.002602	Tellurium	Te	52	127.60
Holmium	Ho	67	164.93032	Terbium	Tb	65	158.92534
Hydrogen	H	1	1.00794	Thallium	Tl	81	204.3833
Indium	In	49	114.818	Thorium*	Th	90	232.0381
Iodine	I	53	126.90447	Thulium	Tm	69	168.93421
Iridium	Ir	77	192.217	Tin	Sn	50	118.710
Iron	Fe	26	55.845	Titanium	Ti	22	47.867
Krypton	Kr	36	83.80	Tungsten	W	74	183.84
Lanthanum	La	57	138.9055	Uranium	U	92	238.0289
Lawrencium*	Lr	103	262.110	Vanadium	V	23	50.9415
Lead	Pb	82	207.2	Xenon	Xe	54	131.29
Lithium	Li	3	6.941	Ytterbium	Yb	70	173.04
Lutetium	Lu	71	174.967	Yttrium	Y	39	88.90585
Magnesium	Mg	12	24.3050	Zinc	Zn	30	65.39
Manganese	Mn	25	54.938049	Zirconium	Zr	40	91.224
Meitnerium	Mt	109	—				

*This element has no stable isotopes. The atomic mass given is that of the isotope with the longest known half-life.

NAMES, FORMULAS AND CHARGES OF COMMON IONS

Positive Ions (Cations)		Negative Ions (Anions)	
1+ Ammonium	NH_4^+	Acetate	$C_2H_3O_2^-$
Copper(I)	Cu^+	Bromate	BrO_3^-
(Cuprous)		Bromide	Br^-
Hydrogen	H^+	Chlorate	ClO_3^-
Potassium	K^+	Chloride	Cl^-
Silver	Ag^+	Chlorite	ClO_2^-
Sodium	Na^+	Cyanide	CN^+
		Fluoride	F^-
2+ Barium	Ba^{2+}	Hydride	H^-
Cadmium	Cd^{2+}	Hydrogen carbonate	HCO_3^-
Calcium	Ca^{2+}	(Bicarbonate)	
Cobalt(II)	Co^{2+}	**1−** Hydrogen sulfate	HSO_4^-
Copper(II)	Cu^{2+}	(Bisulfate)	
(Cupric)		Hydrogen sulfite	HSO_3^-
Iron(II)	Fe^{2+}	(Bisulfite)	
(Ferrous)		Hydroxide	OH^-
Lead(II)	Pb^{2+}	Hypochlorite	ClO^-
Magnesium	Mg^{2+}	Iodate	IO_3^-
Manganese(II)	Mn^{2+}	Iodide	I^-
Mercury(II)	Hg^{2+}	Nitrate	NO_3^-
(Mercuric)		Nitrite	NO_2^-
Nickel(II)	Ni^{2+}	Perchlorate	ClO_4^-
Tin(II)	Sn^{2+}	Permanganate	MnO_4^-
(Stannous)		Thiocyanate	SCN^-
Zinc	Zn^{2+}		
3+ Aluminum	Al^{3+}	Carbonate	CO_3^{2-}
Antimony(III)	Sb^{3+}	Chromate	CrO_4^{2-}
Arsenic(III)	As^{3+}	Dichromate	$Cr_2O_7^{2-}$
Bismuth(III)	Bi^{3+}	Oxalate	$C_2O_4^{2-}$
Chromium(III)	Cr^{3+}	**2−** Oxide	O^{2-}
Iron(III)	Fe^{3+}	Peroxide	O_2^{2-}
(Ferric)		Silicate	SiO_3^{2-}
Titanium(III)	Ti^{3+}	Sulfate	SO_4^{2-}
(Titanous)		Sulfide	S^{2-}
4+ Manganese(IV)	Mn^{4+}	Sulfite	SO_3^{2-}
Tin(IV)	Sn^{4+}		
(Stannic)		Arsenate	AsO_4^{3-}
Titanium(IV)	Ti^{4+}	Borate	BO_3^{3-}
(Titanic)		**3−** Phosphate	PO_4^{3-}
5+ Antimony(V)	Sb^{5+}	Phosphide	P^{3-}
Arsenic(V)	As^{5+}	Phosphite	PO_3^{3-}

CPSIA information can be obtained at www.ICGtesting.com
Printed in the USA
BVOW04n2144280116

434382BV00001B/1/P

9 781118 330890